The Dyeing of Textile Fibers

Theory and Practice

The Dyeing of Textile Fibers

Theory and Practice

Joseph Rivlin

Joseph Rivlin
Professor, Textile Chemistry
Department of Chemistry and Physical Science
Philadelphia College of Textiles and Science
Philadelphia, Pennsylvania 19144

Cover Design: James McDevitt
Cover Woven Fabric: Tiffany DeSouza

ISBN : 0-9633133-0-4

Copyright © 1992 Joseph Rivlin

Preface

This book is aimed at providing a fundamental understanding of the principles underlying the various steps that take place in the complex process of dyeing textiles. The purpose of the book is to present the field of dyeing in all its diversity. Whereas the emphasis is on the wet processing of textiles, the book also conveys the notion that dyeing is a coherent intellectual enterprise. The main goal of the text is to encourage the student to develop a practical but basic understanding of dyeing processes, rather than trying to master the methodology of dyeing through memorization.

The book is written as a teaching text. It is designed both for college students entering the field of textiles, whether it be in production, research, or consumer service, as well as for those in industry interested mainly in the practical aspects of the dyeing of textiles. In order to fully benefit from the book, the student should have introductory courses in General and Organic Chemistry.

In the industrial setting, the text is suitable not only for those involved directly with coloring textiles, but also for all those in the textile or allied industries who need the understanding of the subject to help them in their work.

In this short text on such a diversified subject, it was necessary to concentrate on topics that are of prime importance in understanding basic concepts of coloring and wet processing of textiles. This text combines modern practical aspects of coloring textiles with the essential theoretical background for understanding them. Whenever possible, explanations are given in a simplified form. Usually concepts are developed from first principles according to the old Chinese proverb that states: 'Even the longest trip starts with a single step'.

All procedures and applications described in the text serve as guidelines, and are not specific instructions to be used in specific dyeing procedures.

J.Rivlin

Acknowledgments

In writing this book I have received much help and advice from many individuals. In particular I would like to acknowledge the scientific and editorial help of:
Dr. Mendel Trachtman, Chairman, Department of Chemistry and Physical Science, Philadelphia College of Textiles and Science.
     Special thanks are due to my wife, Yael, and my children Michelle and Ze'ev, who consistently encouraged my efforts and who have been true partners in providing support in many different ways.

Joseph Rivlin
Philadelphia, Pennsylvania
April 1992

CONTENTS

Chapter 1. INTRODUCTION — 1
1.1 Characteristics of Dyes — 3
1.2 Information on Coloring Materials — 5
1.3 Some Physical Properties of Fibers Pertinent to Dyeing — 6
1.4 The Exhaust Dyeing Process — 8
1.5 Effect of Temperature on Exhaust Dyeing — 13
1.6 Dye Absorption Isotherms — 14
1.7 Chemical Structures of Dyes and Fibers — 16

Chapter 2. DETERGENTS AND SCOURING — 17
2.1 Polar, Nonpolar and Ionic Compounds — 17
2.2 The Detergents — 18
2.3 Surface Tension and Surfactants — 19
2.4 Soap — 22
2.5 The Synthetic Detergents (Syndets) — 24
2.6 Classification of Detergents — 24
2.7 The Use of Surfactants in Dyeing — 27
2.8 Water Softening — 27
2.9 Sequestering Agents — 28

Chapter 3. CLASSIFICATION OF DYES AND FIBERS — 30
3.1 Classification According to Solubility — 31
3.2 Classification According to Method of Application — 32
3.3 Classification According to Chemical Constitution — 32
3.4 Fiber Classification — 33
3.5 Physical Forms of Dyes — 35

Chapter 4. COLOR AND CHEMICAL CONSTITUTION — 37
4.1 What is Color? — 37
4.2 Mixing Colors — 40
4.3 The Description of Color and Color Measurement — 43
4.4 Color and Chemical Constitution — 44

4.5 Chemical Structures of Dye Molecules	47
4.6 phthalocyanine Dyes and Pigments	49
4.7 The Cationic (Basic) Dyes	53
4.8 Fluorescent Brightening Agents	54

Chapter 5. DYEING EQUIPMENT — 57

5.1 Dyeing Raw Stock	58
5.2 Dyeing Yarns	59
5.3 Dyeing Fabrics	61
5.4 Equipment for Continuous Dyeing	66

Chapter 6. PREPARATORY OPERATIONS — 72

6.1 Cleaning of Wool	73
6.2 Bleaching	73
6.3 Bleaching with Hydrogen Peroxide	74
6.4 Preparation of Cotton Fabrics	76
6.5 Heat-Setting of Thermoplastic Fabrics	82

Chapter 7. DYEING WOOL WITH ACID DYES — 84

7.1 The Structure of Wool	84
7.2 The Acid Dyes	87
7.3 The Application of Acid Dyes to Wool	88
7.4 The Different Types of Acid Dyes	90
7.5 Acid Leveling Dyes	92
7.6 Acid Milling Dyes	92
7.7 Neutral Dyeing Acid Dyes	93
7.8 Chrome Dyes	94
7.9 Premetallized Acid Dyes (Metal Complex Dyes)	96
7.10 Stripping Acid Dyes From Wool	98

Chapter 8. DYEING CELLULOSE WITH DIRECT DYES — 99

8.1 Cellulose	99
8.2 The Direct Dyes	101
8.3 General Properties of Direct Dyes	103
8.4 The Application of Direct Dyes	104
8.5 Classification of Direct Dyes	107

8.6 After-Treatments of Direct Dyes	108
8.7 Stripping of Direct Dyes	109
8.8 Other Classes of Dyes for Cellulosic Fibers	109

Chapter 9. AZOIC DYES — 111

9.1 General Properties of Azoic Dyes	112
9.2 The Application of Azoic Dyes to Cellulosic Fibers	114
9.3 Printing with Azoic Dyes	117

Chapter 10. VAT DYES — 119

10.1 History	121
10.2 Properties of Vat Dyes	121
10.3 The Application of Vat Dyes	122
10.4 Classification of Vat Dyes	125
10.5 Application of Vat Dyes by Continuous Methods	126
10.6 Soluble Vat Dyes	128

Chapter 11. SULFUR DYES — 131

11.1 The Chemical Nature of Sulfur Dyes	131
11.2 General Properties of Sulfur Dyes	132
11.3 The Application of Sulfur Dyes	133

Chapter 12. REACTIVE DYES — 136

12.1 The Chemical Nature of Reactive Dyes	138
12.2 The Properties of Reactive Dyes	139
12.3 The Application of Reactive Dyes By Exhaust Methods	139
12.4 Cold Pad-Batch Method	142
12.5 Continuous Dyeing with Reactive Dyes	143
12.6 Hetero-Bifunctional Reactive Dyes	144

Chapter 13. DISPERSE DYES — 146

13.1 Commercial Forms of Disperse Dyes	147
13.2 Dyeing Hydrophobic Fibers with Disperse Dyes	147
13.3 Chemical Characteristics of Disperse Dyes	149
13.4 Classification of Disperse Dyes by their Volatility	151
13.5 Dyeing Cellulose Acetates with Disperse Dyes	152

13.6	The Dye-Carriers	153
13.7	Dyeing Acetate Fibers	155
13.8	Dyeing Triacetate Fibers	156

Chapter 14. DYEING POLYESTER — 157

14.1	Heat-Setting of Polyester	158
14.2	Thermo-Migration	159
14.3	Dyeing Procedures for Polyester	159
14.4	Oligomers	160
14.5	Exhaust Dyeing	160
14.6	Atmospheric Dyeing	161
14.7	Pressure Dyeing	163
14.8	Continuous Dyeing by the Thermosol Process	168
14.9	Stripping Disperse Dyes From Polyester	169

Chapter 15. DYEING NYLON FIBERS — 170

15.1	Chemical and Physical Properties of Nylon Fibers	170
15.2	Heat-Setting of Nylon	172
15.3	Dye Selection	172
15.4	Dyeing Nylon with Acid and Metallized Dyes	173
15.5	Dyeing Nylon with Disperse Dyes	175
15.6	Modified Nylon Fibers	175

Chapter 16. DYEING ACRYLIC FIBERS — 177

16.1	Chemical and Physical Properties of Acrylic Fibers	177
16.2	Dyeing Acrylic Fibers with Disperse Dyes	179
16.3	Dyeing Acrylic Fibers with Cationic Dyes	179
16.4	Factors Influencing the Dyeing with Cationic Dyes	182
16.5	Retarders for Cationic Dyes	183

Chapter 17. DYEING POLYESTER/COTTON — 185

17.1	Union and Cross Dyeing	185
17.2	Exhaust Dyeing of Polyester/Cotton Blends	185
17.3	Continuous Dyeing of Polyester/Cotton	187
17.4	Continuous Dyeing Procedures	187
17.5	Dyeing Wool Blends	189

APPENDIX — 193
A. General Precautions — 193
B. Dye-Selection — 194
C. Possible Causes for Inappropriate or Faulty Dyeing — 195
D. Fastness Requirements — 196
E. Relative Size of Dye Molecules — 197
F. Suitability of Dyes for Different Fibers — 197
G. The Ionization of Water and The pH Scale — 198
H. pH Values of Chemicals Used in Wet Processing of Textiles — 199
I. Temperature Conversion — 200

REVIEW QUESTIONS — 201
Introduction — 201
Detergents and Scouring — 201
Classification of dyes and Fibers — 202
Preparatory Operations — 202
Color and Chemical Constitution — 203
Dyeing Wool with Acid Dyes — 204
Dyeing Cellulose with Direct Dyes — 205
Azoic Dyes (Naphthols) — 206
Vat Dyes — 206
Sulfur Dyes — 207
Reactive Dyes — 207
Disperse Dyes — 208
Dyeing Polyester — 208
Dyeing Nylon Fibers — 209
Dyeing Acrylic Fibers — 210
Dyeing Polyester/Cotton Blends — 210
Chemical constitution of Dyes and Pigments — 211

REFERENCES — 213
Books — 213
Periodicals — 214

INDEX — 215

Chapter 1.

INTRODUCTION

The main objectives in the dyeing of textile materials are:
1) color the fibers uniformly, and
2) achieve acceptable durability of the color to further treatments in production and normal use. Fastness of the colored material to sunlight and multiple cleaning is of prime importance. Among other aspects that have to be taken into consideration are: reproducing the required shade from batch to batch, using reasonably priced dyes and dyeing procedures, operating according to ecological requirements, and carrying out the coloring process in the shortest possible time.

The coloring of textile fibers is accomplished by a variety of methods. Fibers can be dyed at various stages of production: i.e., in the form of raw stock (loose fibers), tow, yarns, fabric or garments. Synthetic fibers can also be colored during their formation. In this method, called 'solution dyeing' or 'mass coloration', colored pigments are added directly to the spinning solution or to the polymer melt prior to their extrusion, through a spinneret, into filaments. Though solution dyeing yields colored fibers with excellent fastness properties, because of fast changing fashions it is used to a very limited extent.

The majority of the textile fibers is dyed by conventional batch methods of dyeing, referred to as 'exhaust dyeing'. In a typical exhaust dyeing a concentrated solution or dispersion of the dye(s) in water is made first. Then the dye-bath is prepared by diluting the concentrated dye solution with the proper amount of water. Certain chemicals, used as dyeing assistants, are usually added to the dye-bath before and/or during the dyeing to assist in the proper absorption of the dyes by the fibers. Throughout the dyeing process, the dye solution is circulated through the substrate (raw stock, yarn, or fabric) and/or the substrate is moved within the dye-bath. The temperature is gradually raised usually to temperatures close to the boiling point of water, where it is kept for 30 to 60 min. or more, according to need. While dyeing, the dye is taken up slowly (exhaustion of the dye-bath) by the fibers. Usually almost all of the dye is consumed, and the dye-bath is said to be exhausted. The above procedure may vary greatly from one fiber type to another, and from one class of dyes to another. After absorbing the dye, the fibers are further treated, when needed, to obtain the desired wash-fastness. The dyeing process is completed by washing and/or rinsing to remove unfixed dye and dyeing assistants.

Substantial quantities of fabrics are dyed or printed by continuous methods. In continuous dyeing a concentrated dye solution or dispersion is continuously applied across the full width of the fabric usually by a padder, or by spraying, and sometimes other methods. In a padder, the fabric in its open-width form is passed through the dye-liquor and then squeezed between two heavy rollers under pressure. The purpose of padding under pressure is to bring the dye solution and the fabric in close contact, remove excess dye-solution, and achieve an even application. Fixation of the dyes follows by passing the impregnated fabric through dryers and/or curing ovens and/or steamers. The process is completed by passing the fabric through a set of washers to remove unbonded dyes and dyeing assistants.

Continuous dyeing is the preferred method for dyeing very large quantities of fabric, where dyeing with good overall fastness can be achieved at a relatively low cost. In this method fabrics can be dyed by passing them through the dyeing range at speeds of 100 yards or more per minute. Continuous dyeing, however, is much more economical than exhaust dyeing, only when the volume to be processed for each particular shade is approximately 10,000 yards or more.

In painting houses, furniture, etc., pigments are applied to surfaces along with adhesives or binders. Coloration by surface coating is also taking place in pigment printing and pigment dyeing of textile fabrics. In these applications latex binders which can form extremely fine coats on fabric surface are used, and pigments are mechanically mixed with the latex to provide the desired color. However, in dyeing, the dye in the form of single molecules, actually penetrates into the interior parts of the fibers, fixation being achieved without the use of adhesives. In most dyeing applications chemical bonds are formed between the dye molecule and the fiber, whereas in other applications no real bonding occurs. In the latter situations the dye molecules, once inside the fibers, are rended insoluble or simply trapped mechanically. In other instances combinations of the above take place.

Quality dyeing referred to as 'level dyeing' applies not only to the uniform appearance of the dyed material, but also to the proper distribution of the dye throughout the interior parts of the fibers. Improper diffusion (penetration) of the dye can be quickly tested by untwisting a dyed yarn or pulling apart dyed raw-stock. The presence of lighter-dyed fibers will indicate insufficient penetration. More careful tests can be performed by examining cross-sections of dyed fibers under a microscope. A cross-section of a high-quality dyed fiber appears uniformly colored, showing that the dye penetrated to the center of the fiber. When the dye did not penetrate the fiber properly, the cross-section will have a colored ring on its outer part ('ring-dyeing').

When painting with a bright color over a dark shade, sometimes two layers of paint are required for the proper covering of the old color. However, in dyeing, fibers with dark shades cannot be colored to obtain a bright color unless the dark dye is first removed. The term 'stripping' is used to describe the process of removing color from fibers. The complete stripping of a dye is not a common practice; partial stripping, is done quite often in dye houses on substrates with an excess amount of dye.

1.1 Characteristics of Dyes

Currently all dyes used in coloring textile fibers are synthetic organic compounds. In order for a colored chemical to serve as a useful dye it should have the following features:
(a) intense color
(b) solubility in water
(c) substantivity to the fiber in question
(d) durability to wet treatments
(e) durability to further treatments in production and normal use
(f) safe, easy to handle, and reasonably priced.

(a) The dyes have intense colors (molar absorptivity ε >10,000). Those with bright shades are preferred since duller shades can be made by mixing the brighter ones. Adding large quantities of chemicals or finishes to textile fibers may impair their handle, appearance, etc. Hence, the advantage of using dyes with intense colors is that small amounts of the dyes are sufficient to obtain the desired shades. In practice, the increase in weight of fibers by dyeing is between of 0.1% and 3%. Higher weight gains are sometimes required for dark shades such as black or navy.

(b) Dyes are soluble or dispersible in water. In the latter case, they must have at least some solubility, or have the ability of being converted temporarily into a soluble form.

Water is the vehicle of exhaust dyeing; it carries the dyes to the fibers. Dyes can diffuse into the inner parts of the fibers only in the form of single molecules. Therefore, dyes that form true solutions are in the proper state of single molecules, and are ready to be absorbed by the fibers. These dyes, called soluble dyes, form clear homogeneous solutions, and the individual dye molecules are distributed uniformly throughout the stable solution. Among the soluble dyes are many which tend to form small aggregates, which are made up of a few molecules grouped together. Nevertheless, these aggregates eventually dissociate into single molecules which can then diffuse into fibers.

In contrast, disperse dyes are sparingly soluble in water. They require a surfactant (dispersing agent) to form cloudy dispersions. In dispersion, the disperse dye is in the form of extremely fine particles, each consists of hundreds or thousands of molecules grouped together, capable of staying in water for a limited time only. The solubility of disperse dyes is usually in the range of as little as 0.5 to a few m.g. per liter at room temperature, and their solubility increases with increasing temperature [1]. During the dyeing process, the soluble portion of the dye, which is in the form of single molecules, is absorbed by the fibers. This will cause the solid dispersed particles of the dye to dissolve further, to provide again the 'allowed' amount of single molecules in the dye-bath, and the process of dyeing will continue.

(c) The dyes have to be substantive to the fiber in question. The affinity that a dye has to a particular fiber is referred to by the dyers as 'substantivity'. This affinity is necessary to drive the dye molecules from the dye-bath onto the fibers. The substantivity of a dye to a fiber must be much stronger than its affinity to water. Therefore using dyes with high substantivity will increase the yield of dyeing.

Dyeing is a dynamic reversible process; it is an ongoing movement of dye molecules from the dye-bath to the fibers, from one location on the fibers to another, and back to the dye-bath. During the dyeing process the dye is distributed between the fibers and the dye-bath, and unless the dye is strongly attracted to the fibers it may remain largely in the dye-bath. In order to utilize most of the dye, dyeing conditions are such that the distribution will be greatly in favor of the fibers, so that almost all of the dye(s) will be exhausted toward the end of dyeing.

(d) When required, the dyes must possess the ability to become durable to wet treatments. A high substantivity can sometimes provide sufficient durability of the color to a limited number of mild washings. In many cases, however, an additional after-treatment is needed to obtain the required fastness. One method of fixation consists of treating the dyed fiber chemically so that the dye will become insoluble in water.

(e) Dyes should have fastness to further treatments in production and normal use. For example, durable press finishes usually follow dyeing of cellulose containing fabrics. In this case the dyes should be capable of withstanding the low pH and the high temperatures of the durable press process. In addition, dyes should have adequate fastness to sunlight, bleaching, perspiration, dry cleaning, etc., all according to need.

(f) The dyes should be safe, easy to handle, and reasonably priced.

With this long list of required properties, one would expect that only a small number of dyes would meet all the requirements listed above. However, thousands of different dyes are currently available. Obviously there is not a magic group of dyes that meets all of the requirements. Hence, one has to choose a group of dyes that best compromise the properties necessary for the particular textile product.

1.2 Information on Coloring Materials

Dye manufacturers provide dyers with detailed descriptions of their dyes, regarding their application and behavior in normal use, in the so-called 'color cards' or 'shade cards'. When first applied, the information regarding the application of the dyes should be carefully followed, even by the experienced dyer, to avoid unnecessary complications in production.

The most comprehensive reference on dyes and coloring materials is the Color Index, first published by the Society of Dyers and Colorists (SDC) of England in 1924. The third edition published jointly with the American Association of Textile Chemists and Colorists (AATCC) in 1971, was updated with supplements in 1976, 1982, and 1987 (third edition). Detailed information is given on approximately ten thousand dyes, pigments, and other coloring materials (e.g. food colors), as well as various dyeing assistants. The information includes physical properties, chemical constitution, methods of application, fastness properties of the dyes, and a list of potential fibers to which they can be applied. This data can be located in the Color Index by the commercial name of the coloring material or by its C.I. (Color Index) name. The classification of dyes is basically done by their method of application to textile fibers. In the section containing the disclosed chemical structures of dyes, classification is based on their chemical constitution (e.g. monoazo, disazo, anthraquinone dyes, etc.). Each dye has a C.I. name including the class of dyes it belongs to, its color category, and its specific number (e.g. C.I. Acid Blue 7). Only the following colors (hues) are used for describing a dye: yellow, orange, red, violet, blue, green, brown, and black. The coloring material in question will have included in its name one of the above colors, the one to which its own color is closest. The color Index nomenclature is now used world-wide by the textile industry, its allies, and many other industries involved with the production or use of coloring materials.

1.3 Some Physical Properties of Fibers Pertinent to Dyeing

The fibers 'open up' or swell during the dyeing process to become accessible to dyes. Almost all textile fibers are organic polymers, made up of long chain macromolecules. In some fibers, as in wool, the polymeric chains are slightly branched. There are millions of these polymeric molecules in each fiber, most of which are more or less aligned parallel to the fiber axis. These macromolecules are attached to each other through many weak bonds such as hydrogen bonds and/or Van der Waals forces. Within every fiber there are locations with various levels of orientation or order. The three dimensional structure of the textile fibers is a combination of highly oriented regions, less oriented regions, and disordered regions, all of which are necessary for their proper performance.

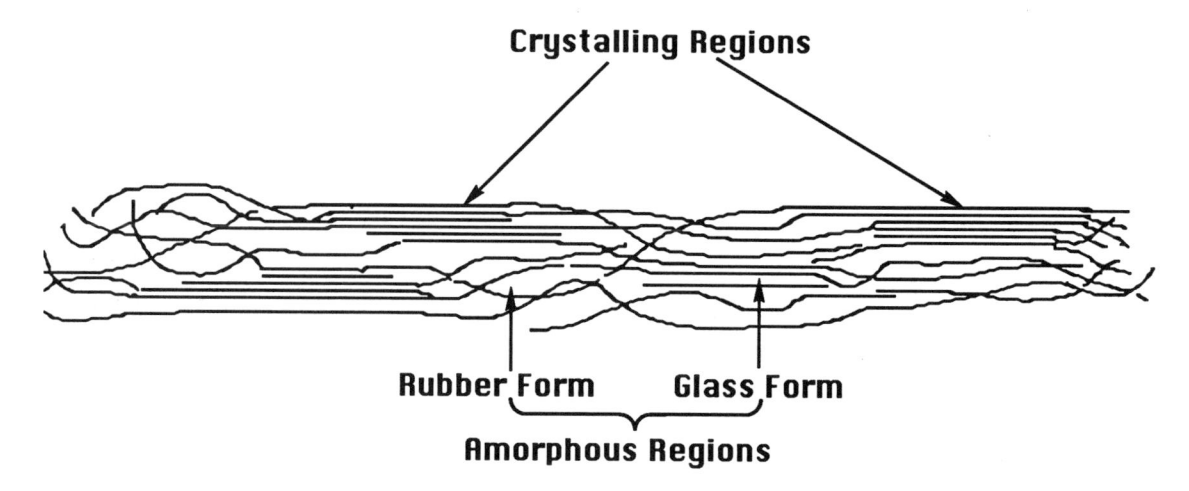

Fig. 1.1 Crystalline and Amorphous Regions in a Fiber

In the highest oriented regions, referred to as crystalline regions (Fig. 1.1), the polymeric chains are aligned parallel and bonded to each other, through a large number of weak bonds, to form a tightly packed network. In the amorphous regions, polymeric chains are randomly distributed, not close enough to form as many bonds with each other as in the crystalline regions, and are therefore freer to move. The crystalline regions are essential for the required strength of the fibers, while the amorphous regions are essential for allowing elasticity, flexibility, and the ability of the fibers to absorb water, dyes, and chemical finishes. X-ray analysis of dyed fibers as compared to undyed fibers show that the distances between the macromolecules in the crystalline regions remain unchanged, indicating that dye molecules did not enter them.

When hydrophilic ('likes water') fibers such as cotton or wool are immersed in water they swell. Water molecules can enter their amorphous regions in between the polymeric chains. As a result, the attractions between the polymeric chains are weakened, and they are forced to move apart slightly from each other. The resulting porous structure (sponge-like) will now allow the larger dye molecules to follow the small water molecules and diffuse into the inner parts of the fibers.

Increasing dyeing temperature also creates more openings in the fiber. With increasing temperature, the movement of molecules increases. In the amorphous regions of fibers, these movements are much more pronounced than in the crystalline regions. The macromolecules in the amorphous regions that are only weakly attached to each other start to move slightly apart from each other with increasing temperature and more openings are created. The higher the temperature the larger are the openings, but the degree of openings or porosity for a given elevated temperature varies greatly from one fiber-type to another.

Hydrophobic ('hates water') fibers such as polyester, do not swell when immersed in water. They do open to dyes, however, as do all fibers, with increasing temperature of dyeing. While the level of opening or porosity in fibers is in most cases rather limited at room temperature, most fibers open up sufficiently at temperatures close to the boil. Some fibers, however, require dyeing at temperatures above the boil, and their dyeing is carried out in sealed dyeing machines under pressure.

Glass Transition Temperatures of Thermoplastic Fibers

In the amorphous (non-crystalline) regions of thermoplastic polymers there are locations with various degrees of orientation. The more oriented ones are rigid and resemble the situation found in glass (Fig. 1.1). Locations within the least amount of orientation consist mainly of randomly distributed coiled polymeric chains with very little attractions to each other, as found in the soft parts of rubber. When the fiber's temperature is raised to its glass transition temperature ($T^o g$), part of its glass form changes into rubber form. This change is referred to as a 'second order transition' which is typical to thermoplastic polymers that do not have a sharp melting point. Dyes can diffuse only into the rubber form of the amorphous region, therefore a significant increase in dyeability is observed when the fiber reaches its $T^o g$. The degree of increase in dyeability due to $T^o g$ varies greatly from one fiber type to the other. For example a very pronounced increase in dyeability is observed when acrylic fibers reach their glass transition temperature. Below their $T^o g$ (approximately 80°C) dye absorption is extremely slow. As soon as the dye-bath temperature reaches the T°g of the acrylic fibers, dyes begin to diffuse into the fibers at a fast rate.

1.4 The Exhaust Dyeing Process

A general summary of the steps involved in exhaust dyeing is given in Fig. 1.2.

Exhaust Dyeing

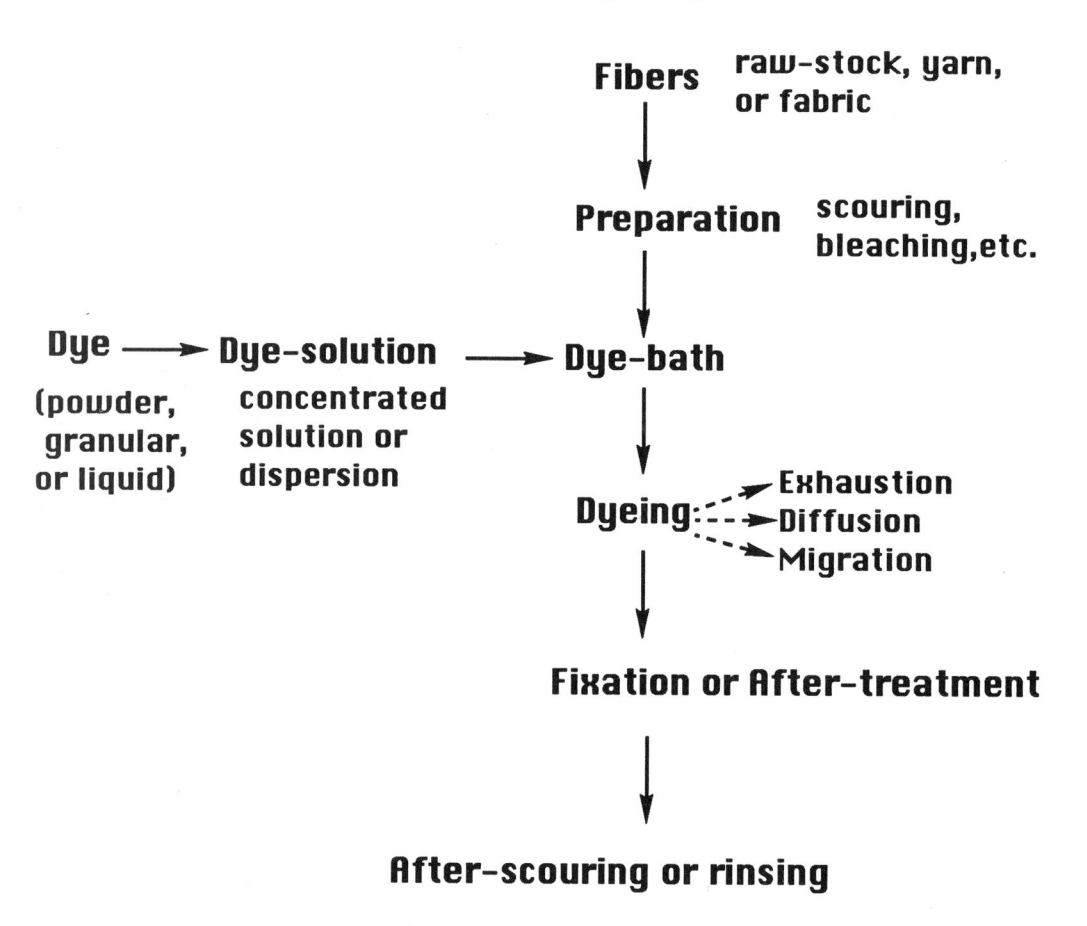

Fibers — raw-stock, yarn, or fabric
↓
Preparation — scouring, bleaching, etc.
↓
Dye (powder, granular, or liquid) → Dye-solution (concentrated solution or dispersion) → Dye-bath
↓
Dyeing → Exhaustion, Diffusion, Migration
↓
Fixation or After-treatment
↓
After-scouring or rinsing

Fig. 1.2 Major Steps in Exhaust Dyeing

Preparation

Prior to dyeing, the textile material is processed to remove all of its impurities that may interfere with the dyeing process, and to prepare the fibers to absorb the dyes properly. Fibers are also treated, when needed, with bleaching agents to remove coloring materials that may prevent reaching the desired shade. Successful dyeing depends not only on the dyeing process itself but also on the preparation of the material prior to dyeing. Even if the dyer is not directly involved in the preparatory operations, he should be well-informed of them prior to dyeing.

Natural fibers contain considerable amounts of impurities. In particular, raw wool may have as much as 70% foreign and natural impurities, such as fats, suint, dirt, and vegetable matter. These impurities should be removed to allow proper diffusion and level dyeing to occur. Dyes used for wool are hydrophilic in nature , and the presence of hydrophobic materials such as fats will repel these dyes and prevent them from penetrating the fibers. Cotton is also dyed with hydrophilic dyes. Raw cotton contains only 1-2% hydrophobic natural impurities; yet this is sufficient to make the fibers difficult to wet and dye. Natural fibers also contain natural pigments which usually make them appear of a yellow shade. These pigments have to be removed by bleaching when bright pastel shades are desired, in particular light blues and violets. In general, all fibers, natural and synthetic, may collect a wide variety of foreign substances and impurities through various processing stages, transportation, and storage. These are impurities such as: grease, lubricating agents, oils, marking dyes, or soil. Best uniform results are obtained with fibers that have been thoroughly and uniformly cleaned and scoured before dyeing.

Inadequate preparation is mostly evident when dyeing fabrics. Woven goods should be thoroughly desized. Other processes such as improper singeing, mercerizing, bleaching, introducing uneven tensions in fabric (barre'), or applying uneven heat-setting, could also lead to unlevelness and faulty dyeing.

Preparing the Dye-Bath

For shade reproducibility, all the amounts of dyes and dyeing assistants and the liquor ratio used should be measured and recorded accurately, in order to be followed precisely in future runs. All dyeing assistants in any dye-bath should be used only if necessary. Using excessive amounts (larger than recommended) of any additive should be avoided.

The rule of thumb is that any dyeing assistant should not be used unless necessary. For example, certain retarders in dye-bath are used to slow down the rate of exhaustion. If, however, heating the dye-bath at a slower rate is sufficient to control exhaustion then it might be better not to use a retarder at all.

The Concentrated Dye Solution

To insure uniformity throughout the dyeing-bath, a small solution or dispersion of the dye(s) is made first. Extreme care must be practiced in the preparation of these concentrated solutions or dispersions. Undissolved or undispersed dye particles, may cause unlevel dyeing or form specks on the goods. Therefore, the dye solution is filtered through a coarse filter before entering the system. Also, when preparing concentrated solutions, dyes and dyeing assistants should not be mixed together in their concentrated form unless specifically required.

To ease the handling and dissolving (or dispersing) of the dyes, the physical forms in which they are supplied have been greatly modified over the years. Today dyes are available in different forms, among them is a liquid form ready for use that needs only to be diluted to the desired concentration.

Form of Substrate

Exhaust dyeing is done on loose fibers (raw-stock, tow, top), yarns (skein, package, warp-beam), fabrics (piece goods), or finished garments. The decision of when and how to dye depends on many factors that should be taken into consideration. Among these are: fashion requirements, economic factors, volume of material to be dyed, fastness requirements, and the availability of specific dyeing equipment.

There are several significant advantages to dyeing raw-stock as compared to dyeing yarns or fabrics. The least expensive methods of obtaining colored fibers by exhaust methods, are those performed prior to spinning them into yarns. Also, dyeing raw-stock is relatively easy, and level dyeing is not critical. This is because the fibers will be thoroughly blended in subsequent operations; in spinning them into yarns and during weaving or knitting. Also, mistakes in matching the required color are relatively easy to correct. For example, a batch of loose fibers too heavily dyed can sometimes be blended with another batch dyed lighter, to obtain the desired shade. Last but not least, the penetration of dyes into loose fibers is significantly faster and sometimes better than the penetration of dyes into yarns or fabrics.

Dyeing raw-stock, however, has a major drawback when used on items that are highly dependent on a fast changing fashion. It may take several months from the time the loose fibers were dyed to the time the finished product is ready for use. During this time fashion requirements may have changed, rendering these products outmoded. In order for the process to be economical, dyeing raw-stock requires coloring large quantities in the same shade. As a result the dye house may be left with large quantities of fibers with unwanted colors.

On the other hand, dyeing fabrics gives the manufacturer the advantage of having more flexibility to satisfy fashion requirements since color decisions are made at a very late stage in production. However, it is much more difficult to achieve level dyeing on fabrics, where uniform results are essential. Faulty dyeing at this stage is very difficult to correct, and such mistakes can be very costly.

Dyeing

For a dyeing procedure to have a commercial value it is extremely important that the application is carried out in the shortest possible time. In summary the rate of dyeing depends on:
1. fibers,
2. dyes, and:
3. dyeing machine.

The way in which the dye moves from the dye solution into the interior parts of the fibers is complex. A great deal of effort has been made by theoretical scientists to study the nature of the dye molecules' movements involved in the dyeing process. Thomas Vickerstaff's book: "The Physical Chemistry of Dyeing" [2] is a corner stone in the development of the theory of dyeing. In his book (p. 123) he wrote: " .. The dyeing must involve three stages, namely:
1. Diffusion of dye through the aqueous dye-bath to the surface of the fibers.
2. Adsorption of dye on the surface of fiber.
3. Diffusion of dye inside the fiber from the surface towards the center...".

The following experiment demonstrates the effect of the first of these three stages on the rate of dyeing. Two fabric samples of the same size are dyed the same way in two separate beakers. Each beaker has the same content, and the only difference in the treatments of the two samples is in the agitation. Sample A is dyed while stirring as required by the dyeing procedure, whereas sample B is dyed without any agitation. To insure proper coverage of sample B with the dye-liquor, it can be kept at the bottom of the beaker with the help of a glass-rod. At the end of the dyeing, sample B will be found dyed lighter than sample A, and most probably sample B will appear unlevel. This experiment shows that the diffusion of the dye through the water to the fiber's surface is slow. The effect of circulation of the dye liquor on the rate of dyeing has been studied in detail [2, p.144]. The results show that increasing the rate of flow of the dye liquor past the fibers increases the rate of dyeing. By agitation, the fibers and the dye liquor are brought into a close contact in a short period of time. Indeed, in dyeing equipment where the contact between the material and the dye liquor is very efficient (e.g. a very fast circulation of the dye-liquor through the substrate), the first stage does not effect the dyeing rate greatly.

The second stage of dyeing, the absorption of the dye from the water-fiber interface, is by far the fastest of the three stages [2,3]. It is the third stage, the diffusion of the dye from the surface of the fibers to their inner parts, that is the slowest. This is the stage on which the rate of dyeing depends mostly. The rate of diffusion depends on several factors such as the substantivity of the dye, its molecular size and shape, the accessibility of the fibers under the dyeing conditions, etc.

From a practical point of view, however, dyers prefer to distinguish in the dyeing process the following three steps:
Exhaustion, Diffusion, and Migration.

Degree of exhaustion refers to the portion of dye absorbed by the fibers, which could be partially inside the fibers, and partially adhering onto the fiber's surface. A very fast exhaustion called a 'fast strike', could lead to unlevel dyeing. This is because it is impossible to bring all the textile material into contact with the dye-bath at the same time, and so parts of the material that will come in contact with the dye-bath first will dye heavier. Therefore, when using dyes with high substantivity, retarders may be used to reduce the rate of exhaustion. The retarder will combine temporarily with a large portion of the dye in the dye-bath, thus reducing the number of dye molecules available for exhaustion. As the dyeing continues at the elevated temperatures, the dye-retarder combination will decompose slowly to supply dye molecules ready for exhaustion. The rate of exhaustion is also reduced when using leveling agents, which increase the solubility of the dyes in the dye-bath. In contrast, dyes with low substantivity may require dyeing assistants that reduce their solubility in water, to promote their exhaustion.

Diffusion is, as said before, usually slower than exhaustion. At this stage the dye molecules penetrate into the inner parts of the fibers 'aiming for dye sites'. Dyes with a high substantivity and/or low solubility will tend to adhere more firmly to the fibers resulting in a slower rate of diffusion.

Migration refers to the movement of the dye from heavily dyed sections to less deeply dyed parts and from fibers back to dye-bath. When unlevel exhaustion occurs, migration becomes necessary for uniform results. To enhance migration dyeing is usually carried out at high temperatures where the fibers are more accessible due to swelling. Certain leveling agents are also used to promote migration. They compete with the dye molecules for the dye sites on the fibers and occupy them temporarily, thus forcing the dye to move on to other dye sites within the fibers.

Completion of the Dyeing Process

After dyeing, if the shade obtained is satisfactory, the dye-bath should be cooled down slowly or dropped while hot. Many dyes may precipitate upon fast cooling, deposit on the goods and cause speck dyeing. With certain fabrics, such as acrylic fabrics, a sudden cooling may lead to creases and distortion of their shape.

Fixation or After-Treatment

This treatment will enhance the wash-fastness of the dye by:
1. improving the bonding between the dye and the fiber, and/or
2. increasing the insolubility of the dye.

After-scouring or Rinsing

This treatment removes unbonded dyes and dyeing assistants from the fibers, and its procedure varies from a mild rinse with warm water to the use of a detergent solution at or near the boil. With certain groups of dyes, a slight change of shade occurs at this stage yielding the true color. Bleeding of dye from a new cloth in the first launderings could be the result of poor rinsing in this final stage of the dyeing process. Further bleeding of the dye in subsequent launderings, however, indicates poor wash-fastness.

1.5 Effect of Temperature on Exhaust Dyeing

When a substrate is dyed under constant conditions dye absorption will increase until a dynamic equilibrium is reached. At equilibrium, no further change in concentration of dye, in fiber or dye-bath, takes place. However, the movement of dye molecules is dynamic, and the amount of dye absorbed is equal to the amount of dye released from fibers during a given time. Equilibrium may be reached in a short time (e.g. during the dyeing period) or may require many days. Reaching equilibrium is not the goal of exhaust dyeing, and in practical dyeing equilibrium is seldom attained. Whenever possible, the dyer will set dyeing conditions so that toward the end of the dyeing almost all of the dye will be absorbed by the fibers.

The effect of dyeing temperature on the rate of exhaustion and the percent exhaustion at equilibrium, for a specific dye is shown in Fig. 1.3.

```
100  │ % Exhaustion
     │                        80°C
     │                        100°C
 75  │
     │
 50  │
     │
     │
  0  └─────────────────────────
              Time
```

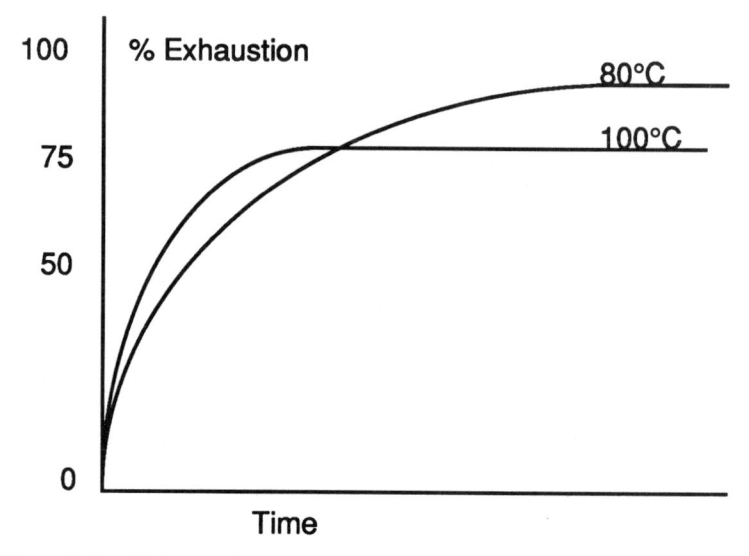

Fig. 1.3 Effect of Temperature on Rate of Dyeing and % Absorption at Equilibrium

In general, as seen in Fig. 1.3, increasing temperature of dyeing increases the rate of dyeing, but decreases the percent exhaustion at equilibrium. This behavior results both from the effect of temperature on the fibers, and also from the increased movement of dye-molecules with increasing temperature. As temperature increases the fibers open up, allowing a faster absorption of dye. However, the increase in temperature will make it easier for the dye to move back into solution. Accordingly, when comparing dyeing at different temperatures the yield of dyeing at the higher temperature can be significantly lower (Fig. 1.3).

1.6 Dye Absorption Isotherms

Practical information can be obtained from dye absorption isotherms. These curves show the effect of dye concentration on the distribution of dye between the fibers and the dye-bath, at equilibrium. Each isotherm curve corresponds to a single temperature and varying amounts of dye used. The three isotherms shown in Fig. 1.4 are typical to those observed in dyeing applications. Isotherm curves are obtained by plotting dye-concentration in fiber (ordinate) against dye-concentration in dye-bath solution (abscissa) at constant temperature.

[D]f = concentration of dye on fibers
[D]s = concentration of dye in dye-bath

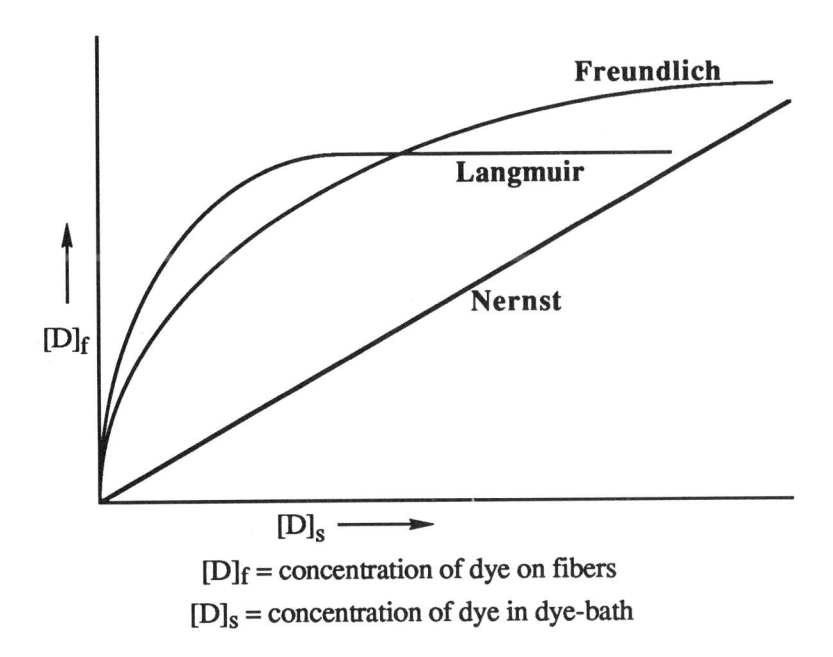

Fig. 1.4   Dye Absorption Isotherms

The Nernst isotherm represents a simple partition of the dye between the fibers and the dye-solution. This curve shows that as dye concentration in dye-bath increases, dye concentration on fibers increases to the same extent. Accordingly : $[D]_f / [D]_s = K$.

The Langmuir isotherm shows that the amount of dye absorbed increases as the concentration increases up to the saturation point. Beyond this point, increasing the dye concentration in the dye-bath will not cause further dye exhaustion. This is typical of fibers with a limited number of dye-sites. As long as there are available dye-sites, absorption will increase with increasing dye concentration in bath, but as soon as all of them are occupied, a further increase in concentration of dye in dye-bath will not increase the amount of dye on fiber. This situation exists, for example, when dyeing nylon with acid dyes, as will be explained in the chapter on the dyeing of nylon.

The Freundlich isotherm shows that as more dye is added to the dye-bath, more dye goes to the fibers. However, as the concentration of dye in the dye-bath increases, the amount of dye on fibers increases less rapidly. No section of the curve is linear. This situation exist when there are sufficient locations available for the dye to attach to the fiber (e.g. dyeing cellulose with direct dyes). At the beginning of dyeing more sites are available, and dyeing takes place at a high rate. As the amount of dye on fibers increases more dye will be attached closer to fibers' surface. This portion of the dye is not as firmly attached to the fibers, thus the amount of dye moving back to the dye-bath increases rapidly.

1.7 Chemical Structures of Dyes and Fibers

The dyer does not have to know the exact chemical structure of the dyes he uses, however, the knowledge of the chemical nature of the dye is essential. This knowledge includes the nature of the chromogen (azo, anthraquinone, etc.), solubility characteristics (anionic, cationic, or non-ionic, type and number of solubilizing groups, etc.), molecular size, volatility, presence of heavy metals, functional groups affecting light and wash-fastness, sensitivity to pH, stability to oxidation and/or reduction, etc. A knowledge of the chemical nature of the dye will enable the dyer to have a better understanding of the dye's behavior during application and the normal use of the finished garment.

A knowledge of certain chemical and physical properties of the fiber is also essential for understanding its behavior during dyeing. These properties include the polymeric nature of the fiber, its functional groups, degree of crystallinity, its reaction with water, effect of temperature on the fiber, thermoplastic characteristics, etc.

References

1. R.H. Peters, Textile Chemistry, Elsevier Publ. Co. Vol. III, 1975, 72
2. Vickerstaff Thomas, The Physical Chemistry of Dyeing, 1954, Interscience Publishers Inc., New York.
3. The Theory of Coloration of Textiles, Second Edition, Edited by Alan Johnson, 1989, Society of Dyers and Colourists., West Yorkshire, England.

Chapter 2.

DETERGENTS AND SCOURING

Water is almost always the medium in which impurities are removed from fibers. It is also used in other preparatory operations, as well as in dyeing and finishing. Therefore it is desirable to be acquainted with the solubilizing properties of water.

The general rule that states: 'like dissolves like' is very useful in predicting the solubility of different materials in a particular solvent. Chemicals made up of polar molecules or ions, tend to dissolve in water, which is a highly polar solvent. Polarity in molecules exists when their bonding electrons are unsymmetrically distributed as in the case of water:

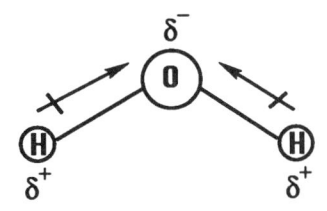

2.1 Polar, Nonpolar and Ionic Compounds

Polar molecules are negatively charged at one end of the molecule, and positively charged at the other end. In water, the oxygen is slightly negative, and the hydrogens are slightly positive. Polar molecules attract each other through their oppositely charged ends, and therefore tend to mix and form solutions. Examples of water-soluble polar compounds are alcohols and carboxylic acids of low molecular weights, and polyalcohols such as ethylene glycol ('anti-freeze'), glycerol, and sugar.

Similarly, chemicals that can form ions, such as salts, acids, and bases, are generally water-soluble. In an aqueous solution of an ionic compound the individual ions are surrounded by water molecules. The negative portion of water molecules are attracted to a positive ion, while the positive ends are directed toward the negative ions. The term hydrophilic is used to describe materials that tend to interact with water.

In non-polar molecules the electric charges are symmetrically distributed; the center of the positive charges and the center of the negative charges are identical and occur at the center of the molecule's mass. Examples of non-polar molecules are: methane (CH_4), and its homologs: ethane, propane, butane, etc. Chemicals made of nonpolar molecules are insoluble in water and are referred to as hydrophobic compounds. Fats, waxes, and grease are also examples of hydrophobic compounds.

Nonpolar or hydrophobic compounds tend to mix with each other, i.e., candle wax in gasoline or octane, or grease in perchloroethylene ($CCl_2=CCl_2$) or trichloroethylene ($CHCl=CCl_2$). The latter two are among the most common nonpolar solvents used in the dry-cleaning of textiles.

2.2 The Detergents

A wide variety of detergents, also called surfactants (SURFace ACTive AgeNTS), are offered for home laundering as well as for preparing textiles for dyeing and finishing. Natural fats, dirt, lubricating oils, waxes, and grease, are among the insoluble materials that must be removed from fibers prior to dyeing. These hydrophobic materials can be emulsified or dispersed in water in the presence of a proper detergent. The term emulsion is used to describe a mixture of finely divided liquid particles (e.g. oil in water), and dispersion is used to describe a fine mixture of a solid in a liquid (e.g. a disperse dye in water). Both, dispersions and emulsions are heterogeneous (non-uniform), appear cloudy and are stable for a limited time only, whereas true solutions (e.g. sugar or soluble dyes in water) are homogeneous (uniform), clear, and stable. The life time of a dispersion or an emulsion may range from a fraction of a minute to years' duration.

The ability of detergents to act as emulsifying or dispersing agents, thus keeping hydrophobic compounds temporarily in water, is the result of their molecular structure. The molecules of a detergent consist of two distinctive components: a hydrophilic part and a hydrophobic part. Thus, detergents can bring together into one mixture two species of an opposite nature.

Common soap is an example of a detergent. It is the sodium salt of stearic acid : $C_{17}H_{35}COONa$

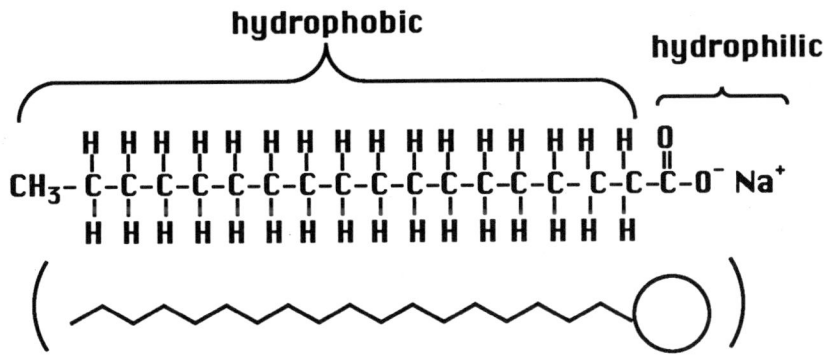

Fig. 2.1 Sodium Stearate (common soap)

In this structure the two parts of opposite nature are: the hydrophobic long hydrocarbon chain repelling water, and the hydrophilic carboxylate group attracting water.

2.3 Surface Tension and Surfactants

The molecules of a homogeneous liquid such as pure water, are attracted to each other through various weak attractions (e.g. hydrogen bonds or Van der Waals forces). Molecules within the bulk of the liquid are attracted by neighboring molecules to the same extent in all directions as shown in Fig. 2.2. Therefore, the net force acting on a molecule in the bulk of a liquid is zero. However, the forces acting on molecules at or near the liquid surface (Fig. 2.2) are unbalanced. These molecules, at the surface, are attracted inward and sideways, but not upward, since they are not completely surrounded by other molecules of the liquid. As a result there is a tension at the surface where the molecules at or near the surface are being pulled inward to the bulk of the liquid.

The surface tension of a liquid may be imagined as a stretched skin covering the liquid. This surface tension is the force that causes a liquid to form spherical droplets when possible, so that the surface area is kept at a minimum. For example, water forms distinctive spherical droplets when spilled on a freshly waxed car. The surface tension of a liquid resists the penetration of other materials into its interior parts. The introduction of a foreign object will increase the surface area of the liquid, and thus increase its surface tension.

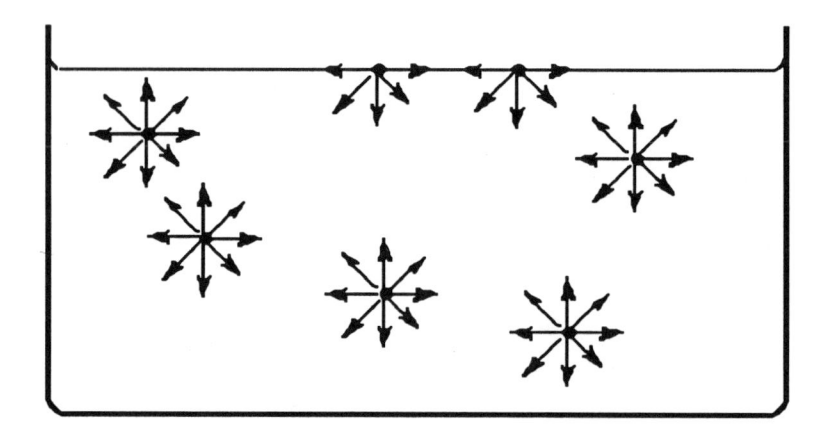

Fig. 2.2 Forces of Attraction on Liquid Molecules

When a detergent such as soap is added to water, it will first concentrate at the surface as an extremely fine film. The soap molecules arrange themselves so that their hydrophilic carboxylate groups stay in the water, while their hydrophobic hydrocarbon chains point outside away from the water as shown in Fig. 2.3.

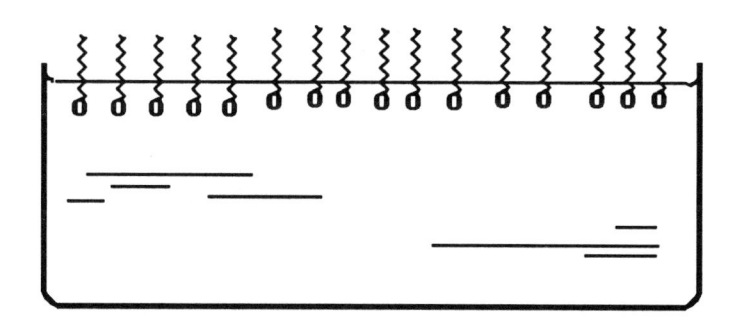

Fig. 2.3 Soap Molecules at the Water Surface

The soap molecules reduce the number of water molecules at the surface, and introduce forces opposing the attraction forces of the water molecules from inside. This results in a significant reduction in the water surface tension. Accordingly, it becomes easier for materials to enter the water and mix with it.

The following experiment demonstrates the strength of the surface tension of water, and the effect of a detergent (surfactant) on it:

A steel needle can remain floating for a while if placed carefully on top of a glass full of water. However, when a tiny drop of a diluted solution of a surfactant is placed carefully at the top of the other end of the glass (away from the needle), the floating needle will sink at once.

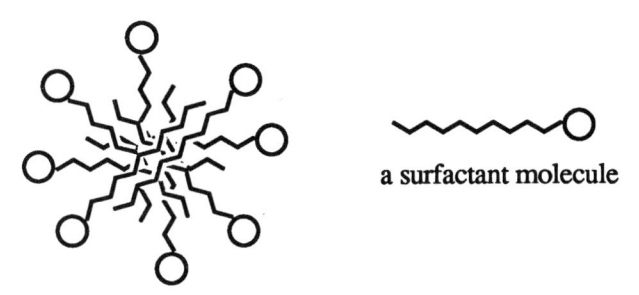

Fig. 2.4 A Micelle of a Surfactant in Water

Further addition of detergent causes 'micelle' formation in the bulk of the water, where the detergent molecules are grouped together to form more or less spherical particles. The molecules are so arranged that the hydrophobic chains form the inside of a micelle, and the hydrophilic ends are at the surface of the micelle orienting themselves outward towards the water. In this way, a micelle as a whole appears as a hydrophilic species (see Fig.2.4), and can remain in water.

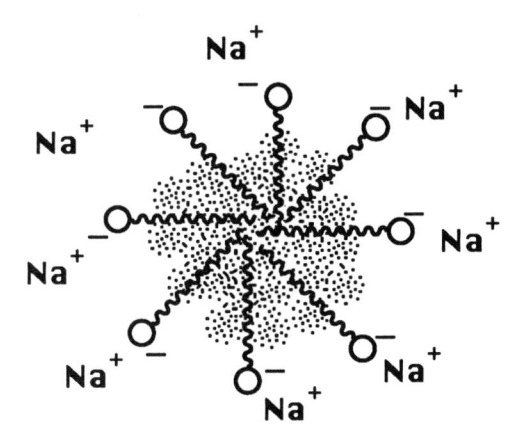

Fig. 2.5 A micelle of an anionic surfactant and a droplet of oil

When emulsions or dispersions in water are made, the extremely fine particles of the hydrophobic material (e.g. very small droplets of oil) become the center of micelles

(Fig. 2.5). The hydrophobic chains of the surfactant are dissolved in the small particles or adhere to their surface. The hydrophilic groups form the surface of the micelle, thus enabling the emulsified material to stay in water

Most of the reduction in surface tension occurs with relatively small amounts of a soap or a detergent. For practical purposes, the concentration of a detergent used in a scouring bath of textiles is in the range of only 0.1%-0.25%. This amount will reduce the surface tension of water from 72 dynes per cm^2 to 30 dynes per cm^2. Additional amounts of the detergent will not decrease the surface tension much further.

Note that whereas all detergents are surfactants, not all surfactants are effective detergents. Many other surfactants are used in wet processing of textiles which are not very effective as cleaning agents. Examples are: certain wetting agents, dyeing assistants, and dispersing agents.

2.4 Soap

The Making of Soap

Soap is the oldest known detergent, its history dates back to the times of the early Romans and even before. The Roman scholar Pliny (A.D.23-79) wrote about a cleaning agent made from goat tallow and wood ashes. This product was a crude form of soap, and its method of preparation resemble in principle to some of the methods used nowadays. Over the years, however, the quality of soap improved greatly in purity and appearance.

Soaps are sodium or potassium salts of fatty acids. The fatty acids are long chain carboxylic acids, and those used in making soaps have between 12 and 18 carbon atoms in the chain. A soap can be made directly by reacting a fatty acid with a proper base:

$$CH_3\text{-}(CH_2)_{13}\text{-}CH_2\text{-}COOH + NaOH \longrightarrow CH_3\text{-}(CH_2)_{13}\text{-}CH_2\text{-}COO^- Na^+$$

Palmitic acid   soap

Note that a soap is a salt made of a weak acid and a strong base. Therefore, soap solutions are basic with a pH of about 10.

Commercial soap is made by treating natural fats or oils with alkali solutions. Fats are esters made of fatty acids and glycerol. The basic hydrolysis of fats, called saponification, yields soap directly:

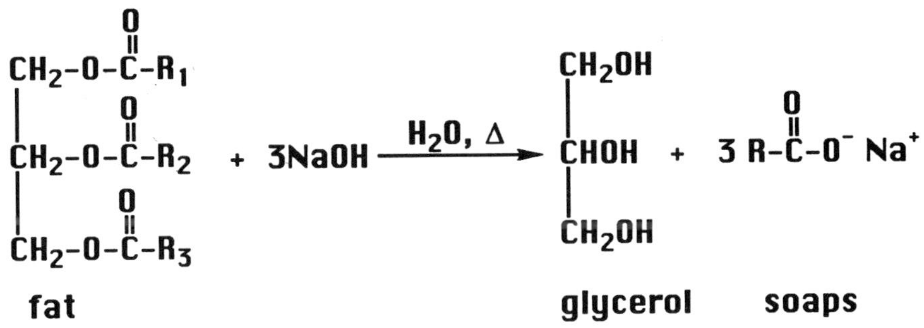

$$\begin{matrix}CH_2-O-\overset{O}{\overset{\|}{C}}-R_1\\ |\\ CH_2-O-\overset{O}{\overset{\|}{C}}-R_2\\ |\\ CH_2-O-\overset{O}{\overset{\|}{C}}-R_3\end{matrix} + 3NaOH \xrightarrow{H_2O, \Delta} \begin{matrix}CH_2OH\\ |\\ CHOH\\ |\\ CH_2OH\end{matrix} + 3\ R-\overset{O}{\overset{\|}{C}}-O^-\ Na^+$$

fat   glycerol   soaps

Fig. 2.6 Basic Hydrolysis of Fats

(R1, R2, and R3, can be the same or different long hydrocarbon chains). The soap obtained in this way has yet to be cleaned from the glycerol, excess alkali, and unreacted fats left in the reaction mixture.

Limitations of Soap

The solubility of soaps in cold water is limited, and warm water is needed to dissolve them properly.

Another disadvantage of soaps is that they form precipitates of fatty acids in acidic solutions. When a strong acid is added to a solution of a salt made of a weak acid and a strong base, the weak acid is liberated. Adding a strong acid to a soap solution will liberate the weak fatty acid(s) which is (are) insoluble in water:

$$CH_3-(CH_2)_n-COO^-\ Na^+ + HCl \longrightarrow CH_3-(CH_2)_n-COOH\downarrow + Na^+Cl^-$$

Sometimes acids are used in the dyeing application itself as for example in the dyeing with acid dyes. If residues of soap are left on the scoured fibers, they may react with the acids in the dye-bath to form insoluble precipitates. Deposits of insoluble fatty acids on the fibers may then lead to unlevel dyeing.

Many times water has a slightly acidic pH (pH5-6) which is caused by industrial pollutants or other impurities. Therefore, in order to insure a basic pH during scouring, soaps as well as other detergents are often mixed with bases. The bases will also react with insoluble acid impurities and convert them into salts which are ionic in nature, and therefore much more soluble in water. Approximate pH values of some chemicals used in wet processing of textiles are shown in Table 2.1, and Table 2.2.

The main disadvantage of soaps is their limitations in hard water. When soap is used in hard water, considerable amounts of card-like precipitates (scum) are formed. These deposits will give scoured fabrics a harsh hand, and if left there they may cause unlevel results in subsequent dyeing applications. It is mainly because of this limitation that the search for new detergents started at the beginning of the present century.

Hard water contains relatively large amounts (more than 100 p.p.m.) of soluble calcium and magnesium salts (e.g. $MgSO_4$, $CaCl_2$). These salts are usually found in considerable quantities in water from wells, rivers, and lakes, that have been traveling long ways through soil and rocks. When soap is added to hard water, Ca and Mg salts of the fatty acids are formed which are insoluble in water:

$$MgCl_2 + 2CH_3(CH_2)_n\text{-}COO\,Na \longrightarrow [CH_3(CH_2)_n\text{-}COO]_2\,Mg + 2NaCl$$

2.5 The Synthetic Detergents (Syndets)

To overcome the limitations of soaps and to obtain additional desirable properties, new surfactants have been synthesized. They are referred to as synthetic detergents because they are usually made from petroleum products, whereas soaps are made mainly from natural fats and oils. The first syndets were synthesized by Reychler, a Belgian chemist, back in 1913, and since then thousands of other surfactants have been developed. Examples of the early syndets are:

sodium lauryl sulfate $CH_3(CH_2)_{10}\text{-}CH_2\text{-}OSO_3^-\;Na^+$, and

sodium dodecylbenzene sulfonate $CH_3\text{-}(CH_2)_{10}\text{-}CH_2\text{-}C_6H_4\text{-}SO_3^-\;Na^+$.

Calcium or Magnesium salts made from these detergents are water-soluble, and therefore, they can be used in presence of hard water without forming precipitates. They can also be used in acidic solutions, though under strongly acidic conditions they may undergo hydrolysis.

2.6 Classification of Detergents

Detergents are classified as ionic or non-ionic depending on whether or not they dissociate into ions when dissolved in water. The ionic detergents are further divided into anionic and cationic detergents, according to the nature of the ion carrying the active part of the detergent.

Note that commercial products that are identified as belonging to one type of detergents, many times, may be a complex mixture of detergents of the same type.

Anionic Detergents
The large majority of the surfactants belong to this group. The active ingredient in this detergents is the negative ion to which the hydrophobic chain is attached. The following are examples of anionic detergents:

R-CH$_2$-COO Na   sodium salt of a long chain carboxylic acid (soaps)

R-CH$_2$-SO$_3$ Na   sodium-alkyl-sulfonate

R-CH$_2$-O-SO$_3$ Na   sodium-alkyl-sulfate

R-CH$_2$-C$_6$H$_5$-SO$_3$ Na   sodium-alkyl-aryl-sulfonate

Also available are anionic surfactants containing more than one ionic site and/or more that one hydrophobic group in their molecular structure.

Cationic Detergents
In these detergents the active ingredients are contained in the cation of the molecule. Many of these are quaternary salts with the general formula:

$$R_1-\overset{\overset{R_2}{|}}{\underset{\underset{R_3}{|}}{N^+}}-R_4 \quad Cl^-$$

where one or two of the R groups is a long chain hydrocarbon group. An example is the following detergent:

$$CH_3-(CH_2)_n-CH_2-\overset{\overset{CH_3}{|}}{\underset{\underset{CH_3}{|}}{N^+}}-CH_3 \quad Cl^-$$

Non-Ionic Detergents
These detergents, as their name indicates, do not dissociate into ions when dissolved in water. The hydrophilic part of a non-ionic detergent is made of a polyethoxy chain:

$$-CH_2-CH_2-O-CH_2-CH_2-O-(CH_2-CH_2-O-)_n CH_2-CH_2-O-CH_2-CH_2-OH$$

This type of a chain contains a large number of polar oxygens that are attracted to water. Since there are no ions in their chemical structures, the non-ionic detergents are practically not affected by hard water. An example of a non-ionic detergent is Triton X 100 made by Rohm and Haas, and has the following structure:

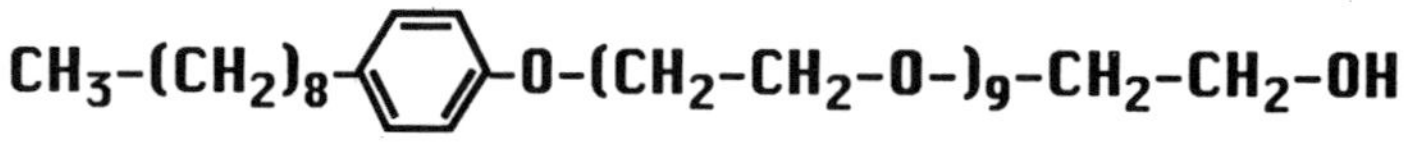

$$CH_3-(CH_2)_8-\underset{}{\bigcirc}-O-(CH_2-CH_2-O-)_9-CH_2-CH_2-OH$$

The Cloud Point

A distinct disadvantage of non-ionic detergents is that their solubility decreases with increasing temperature. The 'cloud point' is the temperature at which the detergent reaches its limiting solubility and start to precipitate out, causing the solution to appear cloudy. In general, when comparing two detergents with similar structures, the one that is more hydrophilic will have a higher cloud point. Though, each detergent has its own cloud-point, when using non-ionic detergents in scouring it is recommended to keep the scouring-bath bellow 70ºC (158ºF). This will avoid attaining the cloud-point of most common non-ionic detergents, and thus avoiding redeposition of oils and other emulsified hydrophobic materials, on the scoured goods.

Amphoteric Detergents

These detergents contain in their active part both a negative charge and a positive charge along with the hydrophobic chain. A general formula for one type of amphoteric detergents is:

$$CH_3-(CH_2)_n-CH_2-\overset{CH_3}{\underset{CH_3}{N^+}}-(CH_2)_n-CH_2-O-SO_3^-$$

Because of their unique structure these detergents can act over a wide range of pH. Sometimes, when mixing anionic and cationic compounds, amphoteric surfactants are used as 'antiprecipitants' to prevent them from combining and settling out of the dye-bath.

Note that even though all of these detergents exhibit the ability to bring together hydrophobic substances and water, many times more than one detergent is needed to prepare a satisfactory emulsion. However, anionic and cationic detergents must not be used together in the same solution. The reason being that, when mixed, the positive ion of the cationic agent may combine with the negative ion of the anionic agent to form an insoluble salt. Even if they remain in solution the complex ions of opposite charges, by being attracted to each other, may cancel each other's activity. However, non-ionic detergents can be mixed with any of the ionic type detergents.

Since the anionic detergents (soaps and sulfuric acid products) were the first to be used as detergents, the cationic detergents are not used in cleaning textile fibers. This is to avoid the possibility of mixing them unintentionally with anionic detergents.

Cationic surfactants, however, are used extensively in textile processing as softeners imparting a 'pleasing hand' to fabrics, as antistatic agents, and as dyeing assistants. A typical softener used in home laundering is a quaternary ammonium salt where: R_1 and R_2 are long hydrocarbon chains, and R_3 and R_4 are methyl groups.

2.7 The Use of Surfactants in Dyeing

Surfactants are used as dyeing assistants for various reasons. They are added to the dye-bath as wetting agents, leveling agents, stabilizers, retarders, antifoams, etc. Surfactants are also present as emulsifiers in chemicals used in the dyeing process, such as lubricants, dye-carriers, etc. In general, the presence of surfactants in the dye-bath increases dye solubility, and therefore dyeing yields may be reduced. Usually non-ionic surfactants have a greater effect on lowering the yield of dyeing as compared to the ionic surfactants. In any case, for successful dyeing the total amount of surfactant in the dye-bath should be kept to the minimum necessary.

2.8 Water Softening

Other wet processing of textiles may also be affected by hard water. Certain chemical finishing agents as well as certain dyes and dyeing assistants may form insoluble salts with Ca^{++} or Mg^{++} ions and precipitate from the application bath. Therefore, it is common practice to avoid hard water in dye houses and other wet processing plants.

Several methods are used to remove hardness from water. The oldest method consists of adding bases such as Na_2CO_3 to the water to form insoluble carbonates of Ca and Mg:

$$Na_2CO_3 + CaCl_2 \longrightarrow CaCO_3\downarrow + 2NaCl$$

The resulting soft water is now filtered to remove the precipitate.

Newer methods involve the use of ion exchangers. The hard water is passed through a cation exchanger (natural and synthetic resins are available) that will replace the Ca^{++} and Mg^{++} ions in the water with Na^+ or K^+ ions. Other transition metals, such as iron and copper ions, though found in water to a much lesser degree, can also interfere with dyeing procedures, and other wet processing applications. These metals are also removed together with Ca^{++} and Mg^{++} in the removal of hardness in water.

2.9 Sequestering Agents

Small amount of undesirable metals present in water can be rendered non active with the aid of a sequestering agent, without removing the metals from the water. Sequestering agents, such as EDTA, HEDTA, or DTPA, form soluble complexes with various polyvalent metal ions (e.g. Ca^{++}, Mg^{++}, Fe^{++}, Cu^{++}, Al^{+++}, Cr^{+++}, etc.) that may interfere with the dyeing (see Fig. 2.7). In the complex, the sequestering agent forms ionic bonds as well as coordinate bonds with the metal. The most common and highly effective sequestering agent is EDTA (Ethylene Diamine Tetra Acetic acid):

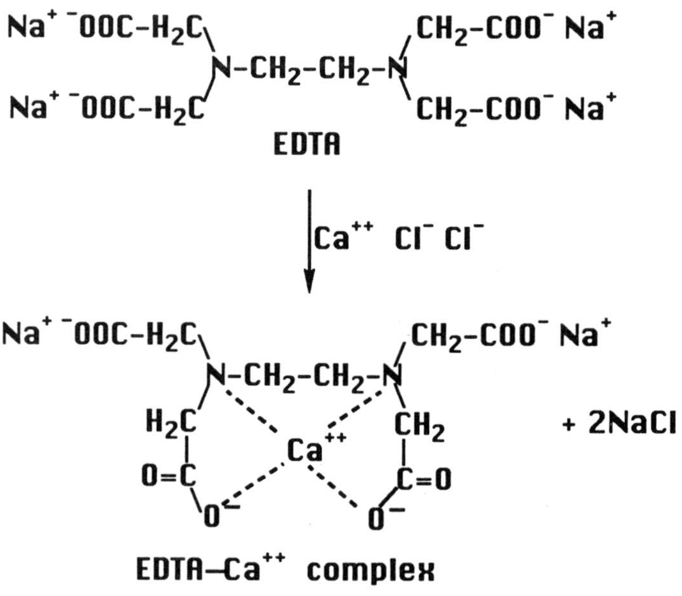

Fig. 2.7 The Reaction of EDTA with a Ca^{++} Ion

Since the complexes formed are made up of rings, sequestering agents are also called Chelating Agents. Note that in the complex there are two carboxylate groups left unattached to the polyvalent metal ion, and therefore the complex as a whole remains water-soluble.

Polyphosphates (Table 2.1) are powerful sequestering agents and are used as such in wet processing of textiles. Since they tend to hydrolyze at high temperatures, however, they should be avoided when dyeing at the boil for a prolonged time.

Table 2.1 pH Values of Phosphates used in Wet Processing of Textiles

compound	Formula	Conc. %w/w	pH (approx.)
Trisodium phosphate (TSP)	$Na_3PO_4 \cdot 12H_2O$	0.1	11.2
"		0.5	11.8
"		1.0	12
Tetra sodium pyrophosphate (TSPP)	$Na_4P_2O_7$	0.05	10
"		1.0	10.2
Sodium tripolyphosphate (STPP)	$Na_5P_3O_{10}$	1.0	9.7
Sodium poly metaphosphate (mixture of oligomers)		0.05	8.3

References

1. Surfactants in Textile Processing, Arved Datyner, Marcel Dekker Inc. Surfactant Science Series, Vol. 14.
2. Detergents and Textile Washing, Principles and Practice, Gunter Jakobi and A. Lohr, 1987, VCH Verlagsgesellschaft, mbH, D-6940 Weinheim, Germany
3. Textile laundering Technology, C.L. Riggs and J.C. Sherrill, 1982, Textile Rental Services Association of America, Hallandale, Florida.

Chapter 3

CLASSIFICATION OF DYES AND FIBERS

There are several different ways by which coloring materials are classified. The following are the most useful ones for the dyer:

<u>1. Classification According to Solubility</u>
Soluble dyes (anionic or cationic)
Disperse dyes (non-ionic, very slightly water soluble)
Pigments (insoluble)

<u>2. Classification According to Method of Application</u>
Acid dyes (protein fibers, polyamides, etc.)
Basic dyes (acrylics, basic dyeable polyester, etc.)
Direct dyes (cellulosics, and some others)
Azoic dyes (cellulosics)
Vat dyes (cellulosics)
Reactive dyes (cellulosics, and some others)
Disperse dyes (polyester, acetates, polyamides, acrylics, and other hydrophobic fibers)

<u>3. Classification According to Chemical Constitution</u>
Azo dyes and pigments (mono azo, diazo, etc.)
Carbonyl dyes and pigments (anthraquinone and indigo derivatives)
Cyanine dyes
Di and triphenylmethane dyes
Phthalocyanine dyes and pigments
and others

3.1 Classification According to Solubility

This is the simplest classification and it is based on the solubility of the coloring materials in water.

The soluble dyes are ionic compounds which dissociate into ions when placed in water. They are hydrophilic and form clear and stable solutions. The disperse dyes are nonionic, hydrophobic, and form dispersions in water. They are made, however, to have a very small solubility in water (at the most a few mg per liter at room temperature) as was mentioned in chapter 1. This very small solubility is essential for the dyeing process, during which the dye enters the fibers only in the smallest possible particle size, namely as single molecules.

Table 3.1 COLORING MATERIALS

PIGMENTS	DYES for exhaust dyeing
Intense color	Intense color
Insoluble in water and common solvents	Must have some solubility in water during the dyeing stage
Are not made to have substantivity to fibers	Must have substantivity to the fibers during the dyeing stage
Molecular size varies from small molecules to polymeric molecules	Molecular size must be small enough to allow the molecules to penetrate the fibers
Stable to further treatments in production and normal use	Stable to further treatments in production and normal use
Durability depends on the binder used	Able to acquire durability to wet treatments

The pigments (Table 3.1) are completely insoluble in water. They are not made to penetrate into textile fibers and therefore are not used in exhaust dyeing. They are used, however, in solution dyeing (mass coloration), and in printing and pigment dyeing along with adhesives (surface coating). The molecular size of a pigment does not have a direct effect on its performance, since it is used in the form of extremely fine particles in the range of 1 micron, consisting of tens to thousands molecules grouped together in one particle. There are pigments of low molecular size, such as indigo, as well as those of polymeric size, such as graphite.

Graphite is used extensively in solution dyeing (mass coloration) and in other applications (e.g. pigment dyeing and pigment printing) as a black pigment. Carbon black is graphite in the form of very small particles in the range of 1-2μ. suitable for solution dyeing. Particles of 10μ or larger in size may clog the holes in the spinneret or weaken fiber filaments. Also, for the same weight of pigment, the smaller the particles, the higher the color yield. Carbon black has excellent fastness properties and very high resistance to reactive chemicals. Since solution dyeing is the least expensive way of coloring fibers and the black color is always in demand, carbon black is used in this method almost as much as all other colored pigments combined.

3.2 Classification According to Method of Application

The classification of dyes by the method in which they are applied to textile fibers is the most common one, and is also used by other industries. The reason for this is that the first synthetic dyes were made for dyeing textile fibers, and it was easy for dyers to call the dyes by names associated with their application. For example the acid dyes are called so because, usually, acids are added to the dye-bath during application. This method of classification will be fully developed, and each group of dyes will be discussed in detail in the following chapters.

3.3 Classification According to Chemical Constitution

This classification is based on the chemical structure of the dyes and the nature of their chromogen. This classification is very useful in predicting some of the properties of dyes relating to their behavior in the dyeing process (solubility, substantivity, etc.) and in normal use (fastness to sunlight, bleaching, etc.).

Note that none of the above classifications is complete by itself. Very often both classifications: Method of Application and Chemical Constitution are given in order to fully describe a dye.

Solvent Dyes

These dyes are hydrophobic and their chemical structure is similar to that of disperse dyes, being mainly mono azo derivatives. They can be applied to hydrophobic fibers from solvents. However, dyeing from solvents is not practical because of the problems associated with pollution and toxicity involved with using large quantities of organic solvents. Currently they are used mainly for coloring petroleum (e.g. gasoline), and other synthetic products.

3.4 Fiber Classification

Table 3.2 Fiber Classification

Hydrophilic Fibers

Protein Fibers (wool, other animal hair, and silk)
Cellulosic Fibers
a. Natural (cotton, linen, jute, etc.)
b. Regenerated (rayon, etc.)

Hydrophobic Fibers

Modified Regenerated (acetate, and triacetate)
Polyester
Polyamides
a. Nylon (nylon 6, nylon 6,6, etc.)
b. Polyaryl amides (nomex, kevlar)
Acrylics
a. Polyacrylonitrile
b. Modacrylic Fibers
Olefins (e.g. polypropylene)
Polyurethanes (spandex)

Textile fibers can be classified according to their chemical constitution, or according to their affinity to water. Chemical structures of the most common textile fibers will be discussed in the following chapters. As to their affinity to water, fibers are classified as hydrophilic or hydrophobic (see Table 3.2). The hydrophilic fibers consist of the cellulosics (cotton, rayon, linen, etc.), and the protein fibers (wool, hair, silk, etc.). These fibers have high moisture regains (Table 3.3), possess very good antistatic characteristics, and swell considerably when placed in water. The synthetic fibers, on the other hand, such as triacetate, polyester, and the acrylics, are hydrophobic. They have low moisture regains and do not swell when placed in water. Nylon, though being hydrophobic in nature, has a moisture regain of about 4% (Table 3.3) due to its open structure, and its hydrophilic amino and carboxyl groups.

Table 3.3 % Moisture Regain of Fibers

Fiber	21°C (70°F), 65% RH
Wool	14-16
Rayon	11-14
Silk	10-11
Cotton	7.5-8.5
Acetate	6-6.5
Nylon	4-4.5
Triacetate	3-3.5
Acrylic	1-2
Polyester	0.4
Polypropylene	0

Fiber-Dye Interactions

As mentioned in the first chapter, in order for a dye to be useful, it has to be substantive (attracted) to the fiber in question. This affinity will enable the dye to be absorbed by the fiber. There are quite a few factors that affect the substantivity of a dye. Among them are the ability of the dye to form strong or weak bonds (e.g. ionic and/or hydrogen bonds and/or Van der Waals forces) with the fiber, its polarity, and its size and shape. The rule of thumb that states: 'like dissolves like', can give us some indication of the expected substantivity. Accordingly, the hydrophobic disperse dyes will dye, under the proper conditions, hydrophobic fibers such as polyester, acetate, and nylon. Similarly the soluble dyes which are hydrophilic in nature are expected to dye hydrophilic fibers. The direct dyes, for example, which are a sub-group of the soluble dyes will dye both cellulosics and protein fibers. However, other soluble dyes are only substantive to certain hydrophilic fibers. Such is the case with the soluble acid dyes which are substantive to proteins but will not dye cellulosics.

3.5 Physical Forms of Dyes

Dyes are mixed with neutral diluents or 'fillers' by the manufacturer to standardize the color strength. A solid sample of a dye may contain as much as 50% of the filler. The fillers are usually chemicals that are used during the application of the dye. For example, sodium sulfate, which is a common filler for direct dyes, is often used in the dyeing process to promote exhaustion. Fillers for disperse dyes are dispersing agents that are needed to stabilize the dispersion.

When the first synthetic dyes reached the market they were supplied as pastes. These pastes were the crude products as obtained from the synthesis of the dye. They had the tendency to dry out, after which it became very difficult to dissolve them. Then dyes were offered in powder forms which were much easier to dissolve. The dry powder, however, being in the form of extremely fine particles, produces during its handling, a 'dust' and escapes into the plant's atmosphere. This dust can reach undyed textiles in plant, or even cause health problems. Certain oils were added to prevent 'dusting', but being insoluble they had the potential of interfering later on with the dyeing process.

Today dyes in other physical forms are also offered, such as the granular, liquid, or paste form. They are easier to dissolve than the powders and do not have the dusting problem. Many dyes are currently sold by dye manufacturers as liquids in the form of highly concentrated solutions or dispersions, usually in concentrations of about 30-50%. These liquids have only to be diluted further, so that the presence of undissolved dye particles in the dye-bath is easy to avoid. Another advantage of the liquid form is that it can be metered automatically, which is an important feature for fully controlled applications. However, liquid forms of dyes have limited stability, and must be kept from freezing and from changing with time.

The following are general procedures for preparing dye solutions (or dispersions) from solids. Soft water should always be used in making these solutions:

Soluble Dyes

This procedure is suitable for acid dyes, and direct dyes. A small amount of water is added to the dye(s), while stirring, to form a smooth paste. Then, boiling water is added and stirring is continued until a clear solution is formed. After cooling, the solution is filtered, if necessary, to remove undissolved dye particles.

Disperse Dyes

Slowly sprinkle the dyes into water (10 to 20 times its weight) at 50°C (122°F), while constantly stirring. After adding all the dyes add warm water (avoid hot water) to the desired volume, and continue stirring for another 10 minutes. This dispersion should then be filtered through a fine-mesh sieve before entering the dye-bath.

Chapter 4.

COLOR AND CHEMICAL CONSTITUTION

4.1 What is Color?

Color is a visual sensation in the brain caused by stimulation of light receptors by visible light energy. Color is seen when visible light reaches the human eye, and strikes the retina at the back of the eye. Light receptors (cones and rods) located on the retina surface absorb visible light and transfer the signal through a complex process which initiates a nerve impulse that is translated in the brain to a visual sensation of color.

Visible light is energy in the form of electromagnetic waves. The x-rays that are used in medicine, the ultraviolet radiation that causes sun burns, and the radio waves that are used for communication, are also energy radiations in the form of electromagnetic waves. They all share certain properties, such as traveling straight in the form of waves at the 'speed of light' (approximately 186,000 miles per second), but differ in their energy content. Those with longer wavelengths have lower energy contents and vice versa. The electromagnetic spectrum extends over a wide range of radiations of different wave lengths as shown in Table 4.1. It includes very powerful radiations of very short wave lengths, such as gamma rays and x rays as well as very weak radiations of very long wave lengths, such as radio waves. The visible region of the electromagnetic spectrum is but a small part of it and ranges from 400 to 750 nm (Table 4.1).

When light composed of all the wavelengths (mixed together approximately to the same extent) within the visible spectrum reach the eye, the color seen is white. The visible part of sunlight at noon is a mixture of the different waves of the visible region, all with approximately the same intensity. Thus, an object appears white at ordinary day light because it reflects the entire visible spectrum and does not absorb any portion of it, whereas an object appears black because it absorbs the entire range of the visible spectrum and no light is reflected An object with a color other than white or black absorbs only part of the visible light spectrum. A colored object appears as it does as a result of the part of the visible light which it reflects. For example a red object at day light appears so because it reflects red wavelengths and absorbs all the others. The key, however, is the wave lengths of the light which are absorbed by the object (Table 4.2). Accordingly a yellow object at day light appears so either because it reflects yellow light (580-595nm) (Table 4.1) and absorbs all other wavelengths, or because it absorbs blue light (435-480nm) and reflects all the others (both green and red [500-700nm]) (Table 4.2).

Table 4.1 The Electromagnetic Spectrum

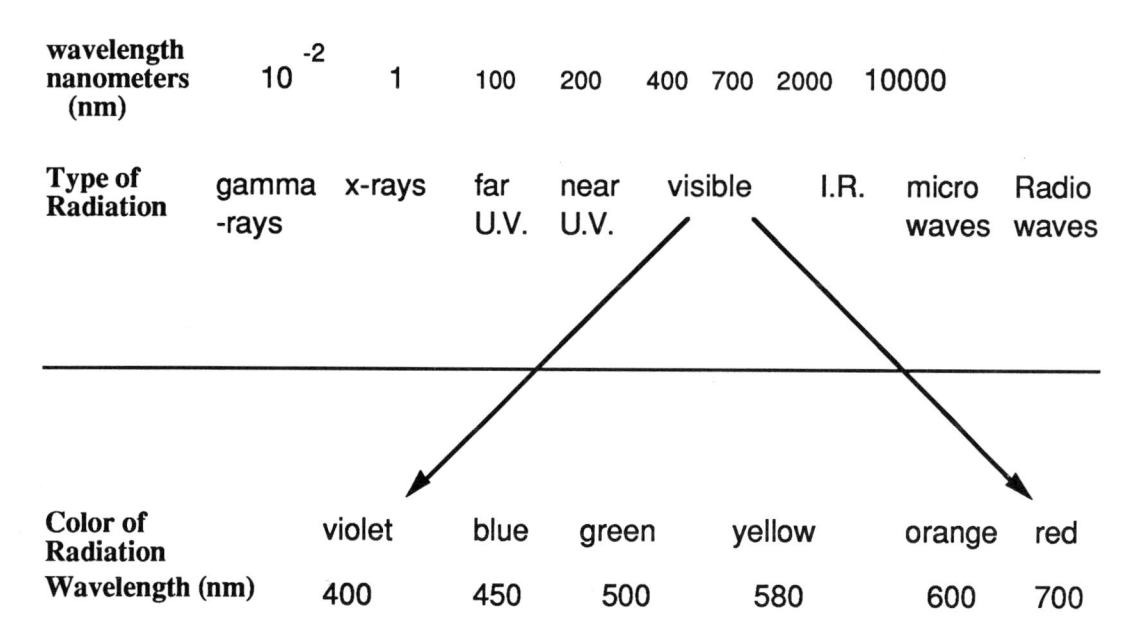

wavelength nanometers (nm)	10^{-2}	1	100	200	400	700	2000	10000
Type of Radiation	gamma-rays	x-rays	far U.V.	near U.V.	visible	I.R.	micro waves	Radio waves

Color of Radiation	violet	blue	green	yellow	orange	red
Wavelength (nm)	400	450	500	580	600	700

The first conformation showing that light is the origin of color came from the experiment in which a prism is used to separate visible light in to its colored light components. When a narrow beam of white light (e.g. sunlight) is passed through a glass prism it is separated into its colored light components. This ability of the prism to separate visible light into color bands was already known to the ancient Greeks (3rd century B.C.). However, they believed that there was a material inside the prism that colored the white light as it passed through. It was only in the beginning of the eighteenth century that Newton explained this phenomenon as we understand it today. He proved it simply by placing a second prism next to the first in such a way that the separated colored bands were combined back to white light. This idea, that white is indeed a color, or in fact a mixture of all the visible spectrum colors, is against the nature of human feeling. It was because of this that a hundred years later (~1780) the great German writer and scientist Goethe refused to accept Newton's explanation. He argued that white is the brightest of all and could not possibly consist of darker elements (colored lights). He also claimed that white is fundamental in nature and cannot be broken down further.

Table 4.2 The relationship between wavelengths absorbed and color perceived

Light Absorbed by the Object	*Wavelengths (nm) Absorbed (λmax)	Color Perceived
violet	400-435	yellow-green
blue	435-480	yellow
greenish-blue	480-490	orange
bluish-green	490-500	red
green	500-560	purple
yellow-green	560-580	violet
yellow	580-595	blue
orange	595-605	greenish-blue
red	605-700	bluish-green

* Range of maximum absorption

The visible light radiations of sunlight at noon have similar intensities. When a beam of sunlight is passed through a prism a continuous spectrum is obtained, where the colors blend one into another continuously as in a rain bow. The colors that are seen are: violet, indigo, blue, green, yellow, orange, and red. Artificial sources of radiation, such as tungsten lamps, fluorescent tubes, etc., contain wave lengths with different intensities. Yet other sources of light, such as sodium arcs and mercury lamps have only a few isolated color bands. When their spectrum is separated by a prism, a few colored lines are seen separated by dark areas. Accordingly, the same object may look different under different sources of light. For example a red car would appear black (or gray) under an all-yellow sodium vapor street light. The red color of the car is not seen since there is no red light in the sodium vapor light; only yellow light is radiated and this light is absorbed by the car.

4.2 Mixing Colors

The entire visible spectrum can be effectively represented by its three main components: blue and red radiations at the extremes and green at the center. Each of the three represents about one third of the visible spectrum. By mixing different amounts of these three lights almost all colors can be produced. Note that a mixture of lights reflected (or radiated) from the same spot is seen as one color. Unlike the ear which is capable of distinguishing individual notes, i.e. in a musical cord, the eye is not capable of analyzing a mixture of wavelengths. Therefore, the visual sensation of a mixture of colors will be seen as a single unique color (Table 4.3).

Table 4.3 The Relationship between light entering the eye and color seen

Color of Object (Color Seen)	Light Reflected or Radiated from Object		
White	Blue	Green	Red
Black	---	---	---
Blue	Blue	---	---
Green	---	Green	---
Red	---	---	Red
Cyan	Blue	Green	---
Yellow	---	Green	Red
Magenta	Blue	---	Red

There are two different types of color-mixing. One type, which is an 'additive' process, is the mixing of colored lights (Fig. 4.1) . The other type, a 'subtractive' process is the mixing of colored pigments or paints (Table 4.4).

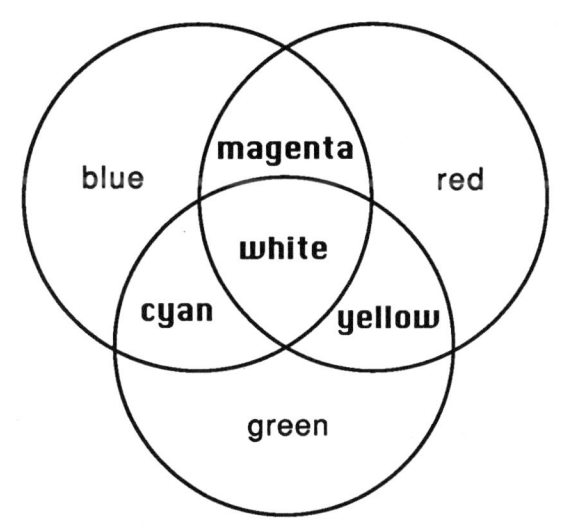

Fig. 4.1 The additive primaries (blue, green, and red lights) projected on a white screen.

The Additive Process of Mixing Colors

The color TV is a good example where the additive process of mixing colors occurs. Here the light comes directly from its source to the eye of the observer. Three colored lights: blue, green, and red, can be radiated from each of the hundreds locations on the screen. These are the three primaries of the additive process, and by mixing different amounts of them, other colors are produced. For example, a yellow color is seen at a certain location when its green and red lights are on. Fig. 4.1 shows the three additive lights: blue, red, and green simultaneously projected on a white screen. As more of the primary lights is added the resulting color becomes brighter and brighter, and when the three lights appear with the same intensity at the same spot, the color white is observed.

The Subtractive Process of Mixing Colors

The three primaries here are: cyan, yellow and magenta. By proper mixing pigments or dyes of these colors, most of the colors that the dyer is interested in, can be produced (Table 4.4). Note that in order to produce the largest possible number of color combinations, the primaries of the subtractive process are as bright as possible and by mixing them duller colors are produced.

Table 4.4 Mixing Pigments or Dyes
(subtractive mixing)

Pigments Mixed	Color Seen
Cyan + Yellow	Green
Cyan + Magenta	Blue
Yellow + Magenta	Red
Cyan + Magenta + Yellow (all to the same amount)	Gray Black
Cyan + Magenta + small amount of Yellow	Navy-blue
Yellow + Cyan + small amount of Magenta	Olive
Yellow + small amounts of Cyan and Magenta	Brown

Color Matching

To simplify color matching and improve reproducibility, the color of at least one of the chosen dyes should be as close as possible to the desired shade. Although it would be advantageous to have as many dyes as possible for a specific application, convenient inventory control restricts the number of dyes which can be kept at hand. Several of the dyes in stock should have colors as close as possible to the required shades as well as to current fashion trends. Some of the dyes in stock should have primary and secondary colors for matching purposes.

In order to produce dull fibers, delustering agents (e.g. titanium dioxide) are added to the fiber solution or the molten polymer. The presence of a delustering agent in the fiber increases the scattering of light, thus increasing the opacity of the fiber. Therefore, the amount of dye required for a given shade increases with increasing the amount of the delustering agent in the fiber.

The sensation of a specific color to the human eye may be brought about by various combinations of different wavelengths. Therefore, it is quite possible that what might appear as a color match under one source of light will appear noticeably different under a different light source. Colors that appear to be the same under one source of light but look different under another light source are called 'metameric' colors. Therefore, the dyer when trying to match a given color using various combinations of dyes, must take the possibility of metamerism into consideration. To avoid metamerism, color matching is examined under several different light-sources.

4.3 The Description of Color and Color Measurement

To describe a color in detail is not an easy task [2, 3]. Several methods for classifying colors have been developed. One of the earliest and commonly used today is the Munsell System [1]. In the Munsell system a huge collection of painted samples is arranged by describing every color by its three attributes: Hue, Value, and Chroma.

Hue is the attribute of color by means of which it is described as being yellow, blue, red, etc. The following colors are used in the Munsell system: violet, violet-blue, blue, blue-green, green, green-yellow, yellow, yellow-red, red, and red-violet. Each of these hues is further divided into ten subdivisions.

Value expresses the lightness or darkness of the color. This is the characteristic that describes the color as lighter (brighter) or darker (duller). Value is expressed by a number 2 to 9 on a scale of 1 to 10 where 1 is an ideal black, and 10 is an ideal white. Thus, when two colors have the same hue but differ in their value number, the one with a higher value number is brighter (Fig. 4.2)

Chroma is the strength or purity of the color. This is the characteristic that indicates the saturation of the color or the amount of color (hue) in the object compared to a gray color with the same value. Accordingly black, gray, and white have zero chroma. Chroma is expressed by numbers of 1 to 12, where a higher number corresponds to a color with a higher saturation.

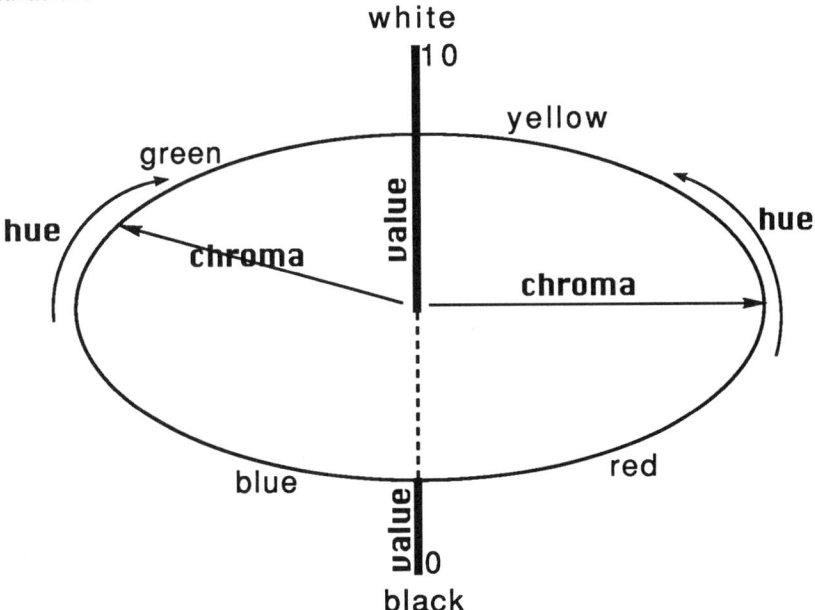

Fig. 4.2 Three Dimensional Color System

In 1931 the Commission Internationale de l'Eclairage (C.I.E.) developed a standard colorimetric method, accepted universally, in which colors are described by numbers [4]. The CIE system is based on the following factors: the standard light source, the reflected light of the object, and the observer. The colored object is placed in a spectrophotometer and illuminated by a standardized light, and the reflected light is measured along the visible spectrum (400 to 700 nm). The products of these three factors across the visible spectrum define the color of an object numerically. One of the major drawbacks of the CIE system is that the colors are not uniformly spaced. In the Munsell system, for example, for any given hue, the chroma scale represents equal visual spacing. To overcome this problem, in 1976 the C.I.E. produced the L*a*b* color space. For a more detailed description using instrumental color matching, where each and every color is defined by a unique set of numbers, see references [2] and [3].

4.4 Color and Chemical Constitution

As mentioned above a substance that absorbs a portion of the visible light appears colored. Various radiations of electromagnetic waves (visible, U.V., IR, X-rays, etc.) can be absorbed selectively by various materials. Molecules will absorb only radiations of definitive wavelengths that can cause some change within their atomic structure. Strong radiations (e.g. U.V.) can excite certain molecules to initiate chemical changes (e.g. photochemical reactions). Absorption of weak radiation, such as visible light, may result in an initial excitation of electrons within the molecule. In most cases, however, the energy absorbed is then converted into heat-energy, and the molecule returns to its normal state.

Visible light can excite electrons in the outer shells of certain molecules such as pigment and dye molecules. The absorbed energy will cause these particular electrons (π electrons) to move from their ground state orbital to higher energy-level orbitals, located further away from their nucleus. The absorbed energy is then converted into heat-energy, and the electrons return to their normal state.

The absorptions of U.V. and visible light are usually in groups of bands. An example of such a band appears in Fig. 4.3. This type of a spectrum is conveniently described by its λmax and ε both of which are characteristic of the absorbing substance. λmax is the wavelength of the tip of the absorption curve. ε is the molar absorptivity, also referred to as molar extinction coefficient, which is a measure of the strength or intensity of the absorption. For compounds absorbing visible light the larger their molar absorptivity, the higher the intensity of the color..

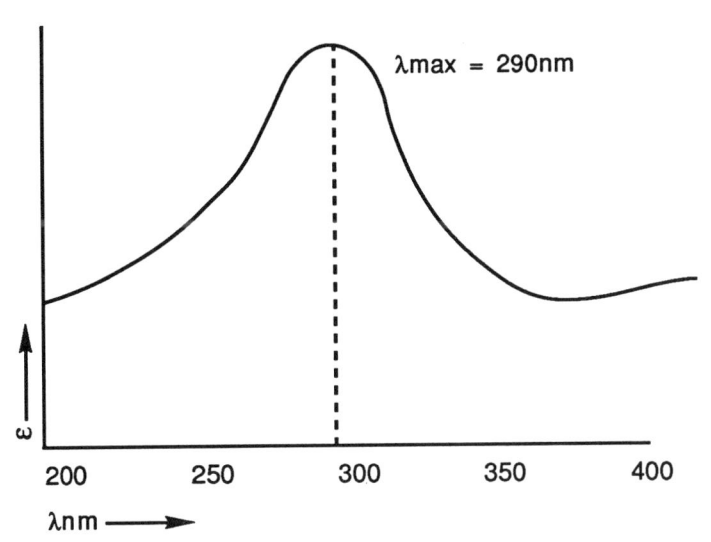

Fig. 4.3 Absorption Spectra of a Compound with Maximum Absorption at 290nm

Electrons involved in single bonds: i.e. C-C and C-H, in saturated hydrocarbons, are held firmly between the atoms and can not be excited by visible light. However, these bonds can absorb strong radiations such as those of the far U.V. in the range of ~130 nm. Unsaturated hydrocarbons absorb radiations of longer wavelength (~180 nm and above) due to the π (pi) electrons of the double bonds which require less energy for excitation. In compounds with conjugated double bonds, such as butadiene or benzene, the π electrons are delocalized and can move within all the carbons (sp^2 carbons) of these systems. Delocalized electrons require less energy for excitation and can absorb weaker radiations with a high intensity. Butadiene, for example, absorbs at 217 nm with an intensity of ε=20,000. By increasing the number of double bonds in the conjugation, the absorptions are shifted further to longer wavelengths (Table 4.5). Thus, beta carotene, the main pigment in carrots, with eleven conjugated double bonds in an aliphatic structure, absorbs visible light with a maxima at ~450 nm (ε=150,000) and appears orange.

To demonstrate the relation between visible light absorption and chemical constitution, it is interesting to compare the two allotropes of carbon: diamond, and graphite. Both are made solely of carbon, and while diamond is colorless and transfers (or reflects) all the visible light, graphite absorbs the entire spectrum of the visible light and therefore appears black. The reason for this is the difference in the way the carbon atoms are bonded to each other in these compounds. In diamond, all the carbon atoms are bonded to each other through single bonds (sp^3 hybridization) to form one gigantic crystal molecule. Therefore, diamond cannot absorb visible light and appears colorless. Graphite, on the other hand, is made of sheets of huge planar molecules consisting of thousands of fused benzene rings (Fig. 4.4).

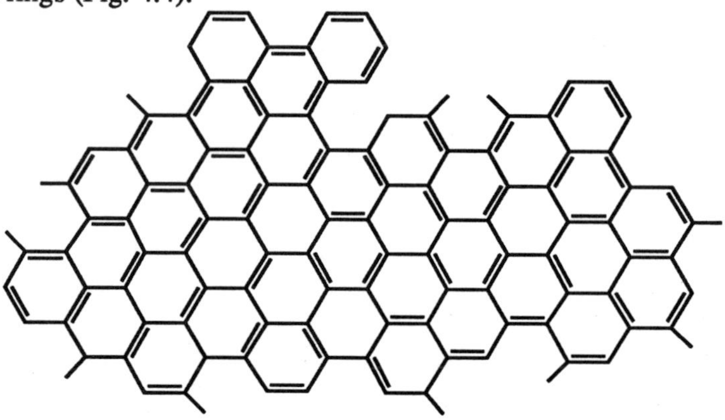

Fig. 4.4 Graphite

Each of the carbons exhibits double bond characteristics (all carbons are with sp^2 hybridization) and the whole macro-molecule contains thousands of conjugated double bonds. Accordingly, the π electrons are highly delocalized and require low energy to be excited, thus molecules of graphite absorb the whole visible spectrum. The delocalization of the π electrons in graphite accounts also for its ability to conduct electricity through its mobile π electrons.

The effect of increasing the number of conjugated double bonds on the color of the molecule is shown in Table 4.5. Every double bond added to the conjugation causes a shift in the absorption toward longer wavelengths

Table 4.5 Colors of Diphenylpolyenes
$C_6H_5(CH=CH)_nC_6H_5$

Value of n	Color
1	none
2	none
3	pale yellow
4	greenish yellow
5	orange
6	brownish orange
7	copper-bronze
11	violet-black
15	greenish black

4.5 Chemical Structures of Dye Molecules

Graebe and Liebermann (1868) were the first to observe that dye molecules contain in their structure conjugated double bonds. A few years later, O. N. Witt perceived that dye molecules contain certain functional groups attached to the conjugated double bonds, which he called 'chromophores', and the combination of the two is the cause for the absorption of visible light. Other functional groups attached to the conjugated double bonds, referred to as 'auxochromes', affect the absorption by shifting it usually toward longer wave lengths and increasing its intensity. The combination of all these three components is the part of the molecule that is responsible for its color, and is called the 'chromogen'.

Currently all dyes are organic aromatic compounds with a conjugated double bonds system, to which chromophores and auxochromes are attached. The presence of these functional groups significantly reduces the number of double bonds in the conjugation, required for intense absorptions of visible light. Accordingly, the resulting dye-molecule will be small enough to easily diffuse into fibers.

Chromophores are functional groups that by themselves absorb visible or near U.V. radiations. They are unsaturated functional groups (except for: $-NR_3^+$) that act as electron acceptors (directing to meta positions in elecrophilic substitution reactions of the benzene ring). Examples of chromophores are:

$-N=N-$	azo group,
$-NO_2$	nitro group,
$-C=O$	carbonyl group,
$-NR_3^+$	alkyl ammonium derivatives, etc.

Auxochromes are saturated functional groups with nonbonding electrons on the atom attached to the conjugated system, and therefore can act as electron donors (directing to ortho-para positions in elecrophilic substitution reactions of the benzene ring). Examples of auxochromes are:

$-NH_2$	amino group,
$-NHR$	mono alkyl amino group,
$-NR_2$	dialkyl amino group,
$-OH$	hydroxy group,
$-OR$	ether group, etc.

A careful examination of chromogens shows that they are made of electron acceptor(s) (chromophore) and electron donor(s) (auxochrome) interacting with each other through a conjugated double bonds system. The affect of attaching an auxochrome to a conjugated double bonds system containing a chromophore is shown below (Table 4.6) with azo-benzene. By adding the dimethyl amino group the strong absorption bands at 330 nm (ε = 17,000) shift to longer wavelengths (λmax 408nm) and a colored molecule with a high intensity (ε = 27,500) is produced. A further shift to longer wavelengths (λmax 478nm) with a further increase in absorption intensity (ε = 33,100) is observed again when another chromophore, the nitro group, is added to the system.

Table 4.6 Absorption Spectrum of Azo Compounds [5]

Compound	λ max nm	ε max L mol^{-1} cm^{-1}
 Azo Benzene	330	17,000
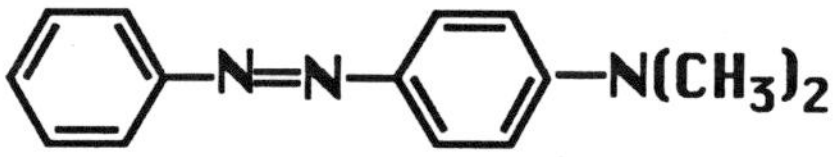 C.I. Solvent Yellow 2	408	27,500
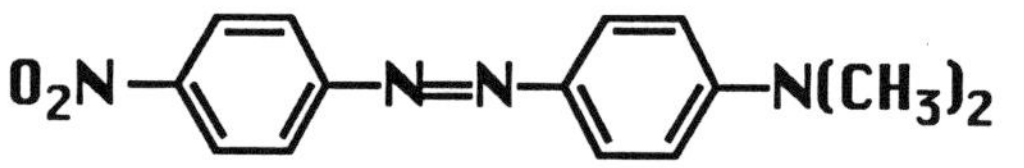 4-Nitro-4-dimethylamino-azobenzene	478	33,100

In addition to the chromogen, dye molecules carry other functional groups, according to need. Soluble dyes have solubilizing groups, usually sulfonic acid (-SO$_3^-$ Na$^+$) group(s) (Fig. 4.5). A reactive dye carries a reactive group(s) that allow(s) the dye to covalently bond to cellulosic fibers. Still other dyes may have a saturated hydrocarbon chain attached to their structure to increase their hydrophobicity or molecular size, etc. An example of a soluble anionic dye is shown below:

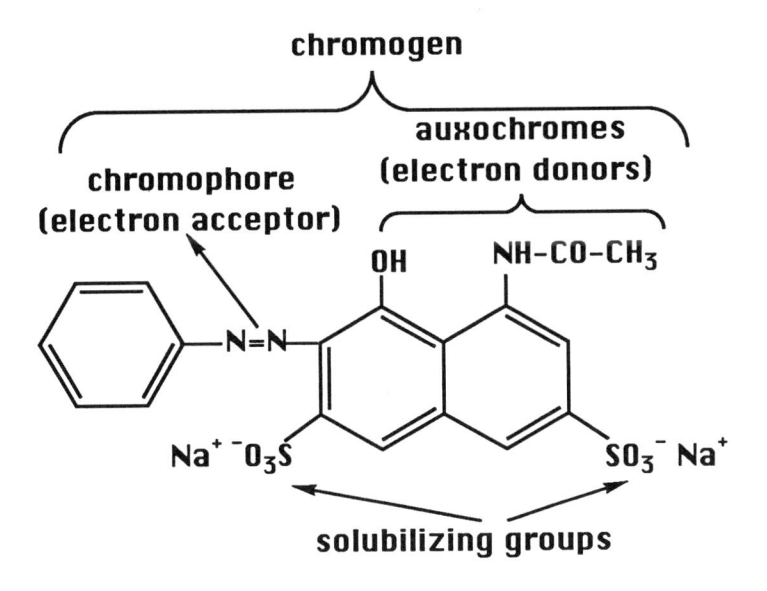

Fig. 4.5 A Soluble Azo Dye   C.I. Acid Red 1

4.6 Phthalocyanine Dyes and Pigments

Very desirable dyes and pigments are those possessing bright intense colors. Usually the higher the color intensity (molar absorptivity) the higher the sensitivity to sunlight. Common types of chromogens are listed in Table 4.7 in order of increasing molar absorptivity. Light fastness of these type of dyes, however, is in the reverse order. Thus, for example, the brilliant triaryl methane dyes are much more sensitive to sunlight than anthraquinone or azo dyes. One of the few exceptions are phthalocyanine (Fig. 4.6) and its derivatives (Fig. 4.7). Cu-phthalocyanine (C.I. Pigment Blue 15), for example, has a very bright blue color, its molar absorptivity in the range of 200,000! and it has an excellent light-fastness!

Table 4.7 Molar absorptivity of Chromogens [6]

Dye Type	Molar Absorptivity (ε)
Anthraquinone	5,000-15,000
Azo	20,000-40-000
Basic cyanine	40,000-80,000
Basic Triarylmethane	40,000-160,000

Phthalocyanine itself (Fig. 4.6) has a very bright greenish-blue color with very good light-fastness. In addition, the phthalocyanines are stable to high temperatures (e.g. heat-setting and pressing temperatures), acids, and bases. The addition of a metal such as Cu, to phthalocyanine significantly increases its light-fastness. Examples of these Cu-complexes are shown in Fig. 4.7 As a result of this stability, Cu-phthalocyanine, which is insoluble in water and most common solvents, is used extensively as a pigment in coloring textile fibers as well as in coloring paints, paper, plastic, etc. It is also used in solution dyeing (mass coloration), and pigment printing and dyeing.

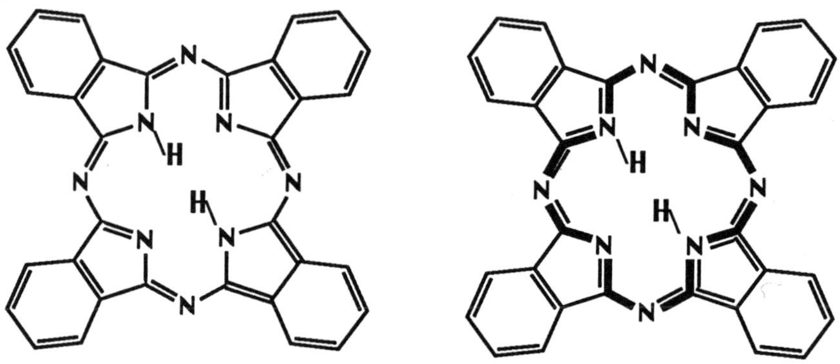

Fig. 4.6 Phthalocyanine

Sulfonation of Cu-phthalocyanine produces water soluble products that can be used as soluble dyes (e.g. C.I. Fig. 4.7, Direct blue 86). The cobalt complex of phthalocyanine (Fig. 4.7, C.I. Vat Blue 29) can be used as a vat dye, and by attaching phthalocyanine to a proper reactive group a reactive dye is produced (e.g.Fig. 4.7, C.I. Reactive Blue 7).

The Chemical Nature of Phthalocyanine
As complex as its structure may seem to be, the synthesis of a phthalocyanine is rather simple. It can be synthesized from common compounds in a single step. Copper phthalocyanine, for example, is obtained by fusing a mixture of phthalonitrile with a copper salt at 200°C. The metal complexes can also be prepared by replacing the two center hydrogen atoms of the metal-free phthalocyanine with atoms such as copper, iron, and cobalt.

The unexpected stability of phthalocyanine and its intense color are the result of its aromatic nature. A cyclic structure with conjugated double bonds is aromatic if it possesses the following properties:
1. All the atoms of the ring should have an available p orbital for the formation of the π orbital of the ring,
2. the number of π electrons in the ring must be equal to 4n+2, where n is an integer (Huckel rule), e.g. for benzene n=1, and
3. the ring should be planar (e.g. all the atoms of the ring exhibit sp^2 hybridization).

A careful examination of the structure of phthalocyanine (Fig. 4. 6) reveals the presence of a 16-membered aromatic ring as shown by the heavy lines in the formula of phthalocyanine. Each of the 16 atoms in the ring (8 carbons and 8 nitrogens) has sp^2 hybridization. The presence of 18π electrons fulfills Huckel rule (n=4), and the entire structure of phthalocyanine has been found to be planar. Hence, its π electrons are highly delocalized and need a low amount of energy to excite them from their ground state to an excited state. Accordingly, phthalocyanines strongly absorbs visible light in the red region (~680 nm), therefore appearing in bluish shades.

The phthalocyanines are structurally related to the chlorophylls (the green pigments of leaves) and to heme (the red pigment of blood). Both have an aromatic ring with 18 π electrons similar to phthalocyanine. However, the superior stability of the phthalocyanine is due to the additional 4 benzene rings (which protect the double bonds that do not participate in the 16-membered aromatic ring), and the absence of aliphatic double bonds as are found in chlorophyll and heme. These benzene rings are not present in the chlorophylls and heme.

There has always been a search to develop phthalocyanine dyes and pigments with different colors. However, the visible-light absorptions caused by the 18-membered aromatic ring are so intense and dominant that substitution on the benzene rings has a very small effect on its color. Presently the only other available phthalocyanine colors of commercial value are a bluish-green and a bright green. These two colors are obtained by polychlorination (12 Cl to 16 Cl) of Cu-phthalocyanine (e.g. C.I. Pigment Green 7), and are used as pigments.

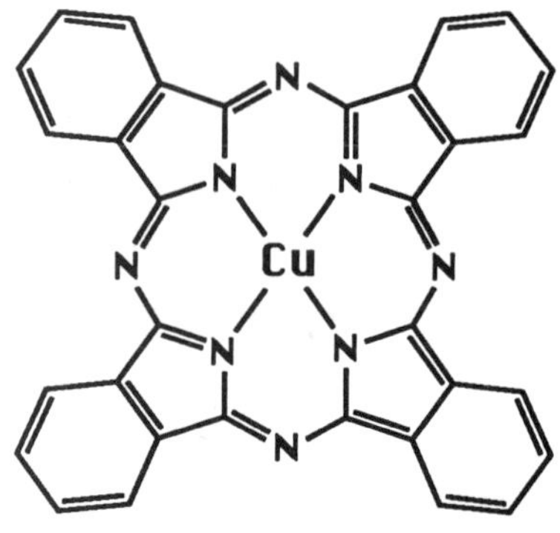

C.I. Pigment Blue 15

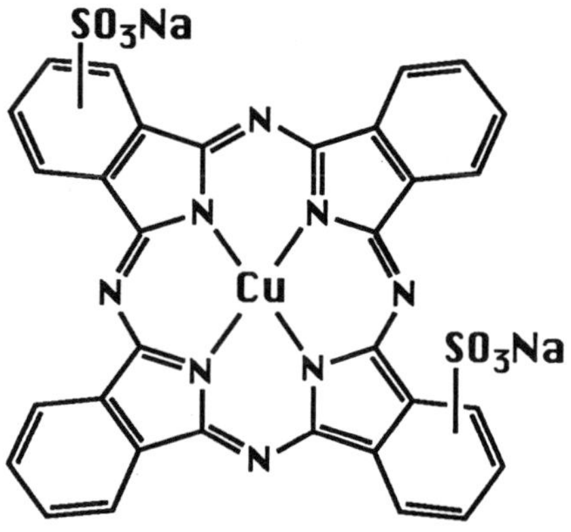

C.I. Direct Blue 86

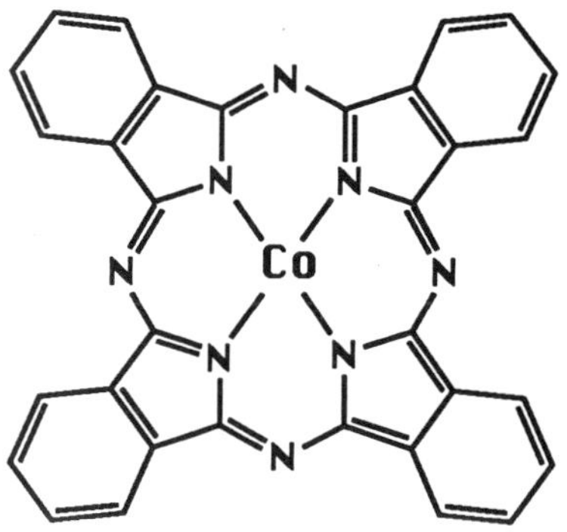

C.I. Vat Blue 29

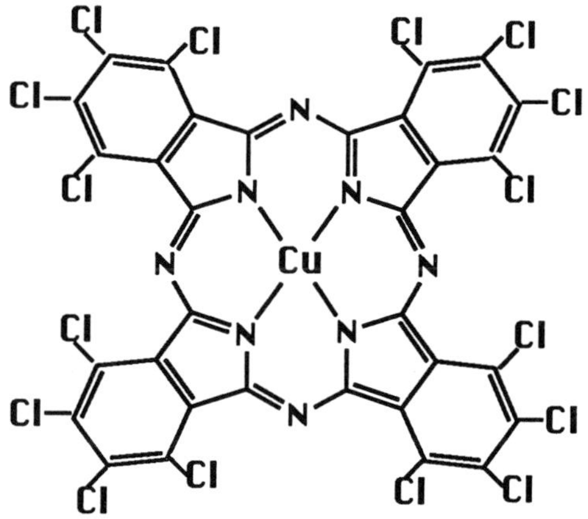

C.I. Pigment Green 7

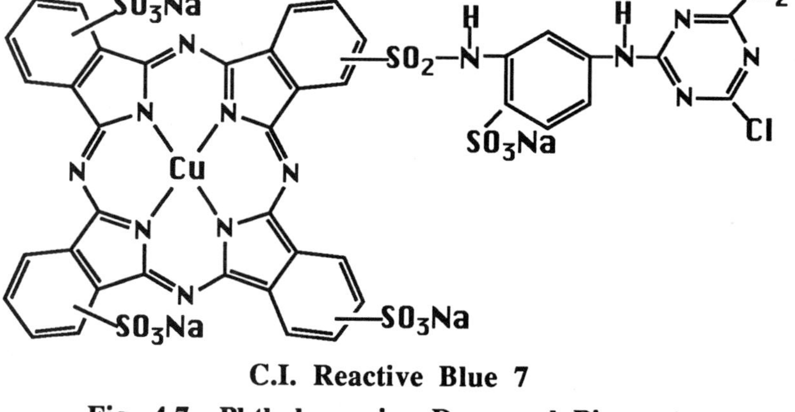

C.I. Reactive Blue 7

Fig. 4.7 Phthalocyanine Dyes and Pigments

4.7 The Cationic (Basic) Dyes

The cationic dyes, also called 'basic dyes', are soluble ionic compounds where the color is contained in the cationic portion of the dye molecule. Most cationic dyes are characterized by their intense and brilliant colors, (some with fluorescent characteristics) but with poor light-fastness. The cationic dyes are available in a complete range of colors, among them brilliant greens. Because of their spectral range and color intensity, basic dyes are considered wherever possible. An example of a cationic dye is:

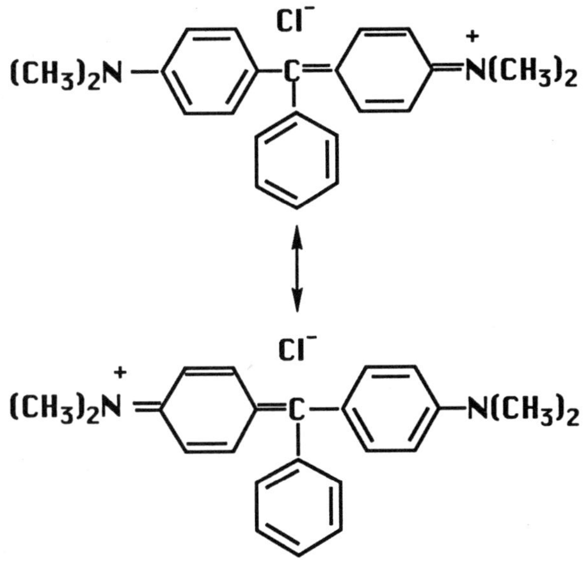

Fig. 4.8 Resonance Structures of C.I. Basic Green 4
(Malachite Green)

Cationic dyes were among the earliest dyes produced synthetically. The first synthetic dye discovered and manufactured in 1856 was the basic dye Mauve.

When first discovered, cationic dyes were applied to natural fibers, the only fibers available at that time. When applied to wool, the colored cations form ionic bonds with carboxylate groups on the fibers. The application to cellulosics involved a pretreatment with tanic acid, as a mordant for bonding the basic dye. Then because of their poor light-fastness, in particular on wool, as well as their limited wash-fastness on natural fibers, the use of cationic dyes on textile fibers was practically discontinued. However, in the fifties, when acrylic fibers were introduced, it was found that when cationic dyes were applied to these new fibers they exhibited excellent wash-fastness, and unexpected durability to sunlight. Other fibers, such as 'basic dyeable nylon' and 'basic dyeable polyester' have been developed to extend the use of cationic dyes on these fibers, and concurrently new cationic dyes more suitable for these new fibers have been developed.

Chemical Nature of Cationic Dyes

Cationic dyes with extremely high intense color have typical chromogens, examples of which are the triarylmethane derivatives, i.e. C.I. Basic Green 4, Fig. 4.8. The extremely high intensity of the color of C.I. Basic Green 4, is the result of the particularly high delocalization of its π electrons. Thus, only a small amount of energy in the form of visible light is sufficient to excite their π electrons. The high delocalization of π electrons in this dye is enhanced by the following factors:

1. the presence of two identical resonance structures (see structure of C.I. Basic Green 4 above), and

2. the presence of functional groups that act simultaneously both as electron donors and electron acceptors, i.e. [-N(CH$_3$)$_2$] and [=N(CH$_3$)$_2^+$]. These two factors contribute largely to a very high degree of delocalization of the π electrons in the conjugated double bonds system of this chromogen.

4.8 Fluorescent Brightening Agents

The fluorescent brightening agents, also called 'optical brightening agents', increase the apparent whiteness or brightness of materials, and are used in textiles, detergents, paper, paints, etc.

Fluorescence is the emission of light caused by radiation. Fluorescent compounds used for whitening have the property of absorbing ultra-violet light and re-emitting energy in the form of weaker energy, i.e. visible light. A part of the absorbed U.V. energy is transformed into heat. Sunlight contains U.V. radiations in the range of 330-380 nm (near U.V.) that are absorbed by optical brighteners used for whitening. The optical brighteners used in textiles re-emit violet-blue light so that the yellow color of the material will appear white. A fabric appears yellow because it absorbs blue light, and reflects the rest of the visible spectrum in the form of a yellow color. When blue light emitted from an optical brightener is added to the reflected yellow light, the color of the fabric will appear white. The whitening produced by optical brighteners is an additive effect, where blue light is added to the reflected yellow light.

The effectiveness of optical brighteners depends on the light source. They operate well at ordinary daylight or under artificial light containing sufficient ultra-violet radiation. Fluorescent light tubes ('day-light'), emit sufficient ultra-violet radiation to obtain an effect similar to that of sunlight. However, optical brighteners are not effective in tungsten light because this light does not contain sufficient ultra-violet radiation.

The Application of Optical Brighteners

Optical brighteners are applied to textile substrates at different stages in production. Insoluble optical brighteners are used as pigments in solution dyeing (mass coloration), where they are added to the spinneret in a manner similar to other pigments. Also available are soluble optical brighteners that are applied to various fibers by exhaust methods.

Chemical Structure of Optical Brighteners

The majority of the brightening agents are stilbene derivatives. An example of a stilbene derivative that is used on cellulosic fibers is C.I. Fluorescent Brightening agent 28:

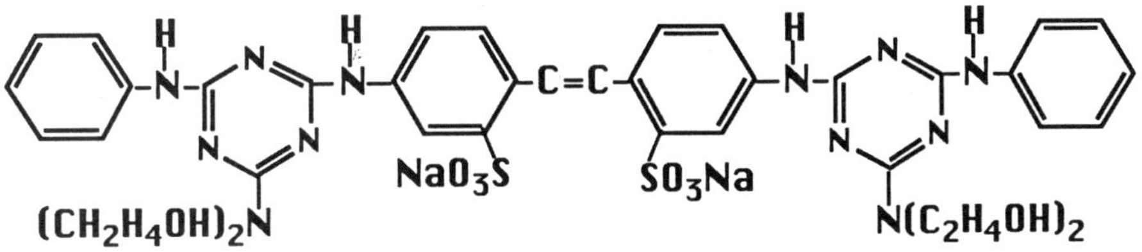

C.I. Fluorescent Brightening agent 28

These types of brighteners have chemical structures similar to that of direct dyes. Accordingly, they can be applied to cellulosics like direct dyes as explained in the chapter on the application of direct dyes to cellulose. However, like direct dyes on cellulosics, their wash-fastness is unsatisfactory.

Note that the above stilbene derivative contains an aliphatic carbon-carbon double bond (C=C outside the aromatic rings) which is sensitive to sunlight, oxidation, weathering, etc. Therefore, these compounds do not have good fastness properties, and tend to loose their ability to absorb U.V. over short periods of time in use.

References

1. the Munsell System (Munsel 1929, 1963, 1969)
2. Color Technology in the Textile Industry, Editors: Gultekin Celikiz and Rolf g. Kuehni, AATCC, 1983, Research Triangle Park, N.C.
3. Principles of Color Technology, 2nd Edition, 1981. Fred W. Billmeyer, Jr., Max Saltzman, John Wiley & Sons, New York.
4. Color Measurment Principles, AATCC Workshop Sept. 23-24,1987, Research Triangle Park, N.C., Chapter 1. Color and the CIE System, Gultekin Celikiz .
5. Fundamentals of the Chemistry and Application of Dyes, P. Rys / H. Zollinger, Wiley-interscience, 1972, p. 11
6. Riegel's Handbook of Industrial Chemistry, Edited by James A. Kent, Van Nostrand Reinhold Co., Chapter 23, D.R. Baer, p. 670.

Chapter 5.

DYEING EQUIPMENT

The main objective of a dyeing machine is to provide maximum contact between the dye-liquor and the fibers within a reasonable amount of time, without causing damage to the material being dyed. The selection of proper dyeing equipment depends on the nature and volume of the material to be dyed. Raw stock and yarns are dyed by exhaust methods, whereas fabrics are dyed both by exhaust or continuous methods. The choice of method for fabrics depends largely on the volume to be dyed. Continuous dyeing is usually considered if the volume of fabric for a particular shade is about 10,000 yards or more.

Many different dyeing machines have been developed for dyeing filaments, loose fibers, yarns, fabrics, and garments. Unlike equipment used in other textile operations, dyeing machines are less standardized, and are many times made to meet special needs in a specific dye-house.

In exhaust dyeing, the contact between the fibrous material and the dye-liquor is achieved by one of the following ways:
1) dye-liquor is circulated continuously by a pump through the fibers that remain stationary, or
2) fibrous material (fabric) is circulated through the stationary dye-liquor, or
3) both are in continuous movement; while the dye-liquor is circulated, the fibrous material (fabric or yarn) is in constant movement.

Modern dyeing equipment are made to operate at low liquor ratios, between 10:1 to 4:1 and lower. The liquor ratio is the ratio between the weight of the dye-bath and the weight of the fibers to be dyed. Low liquor ratios are important because they allow for rapid raising (or lowering) of the dye-bath temperature. In addition to shortening the dyeing time, low liquor ratios are more economical with respect to energy and chemical consumption. However, dyeing at low liquor ratio limits the use of dyes with low solubility.

5.1 Dyeing Raw Stock

Loose fibers are dyed by circulating the dye-solution continuously through the fibers. A typical machine in common use for dyeing loose fibers is the conical pan dyeing machine. These machines consist of a conical pan that fits properly into an outer container (Fig. 5.1).

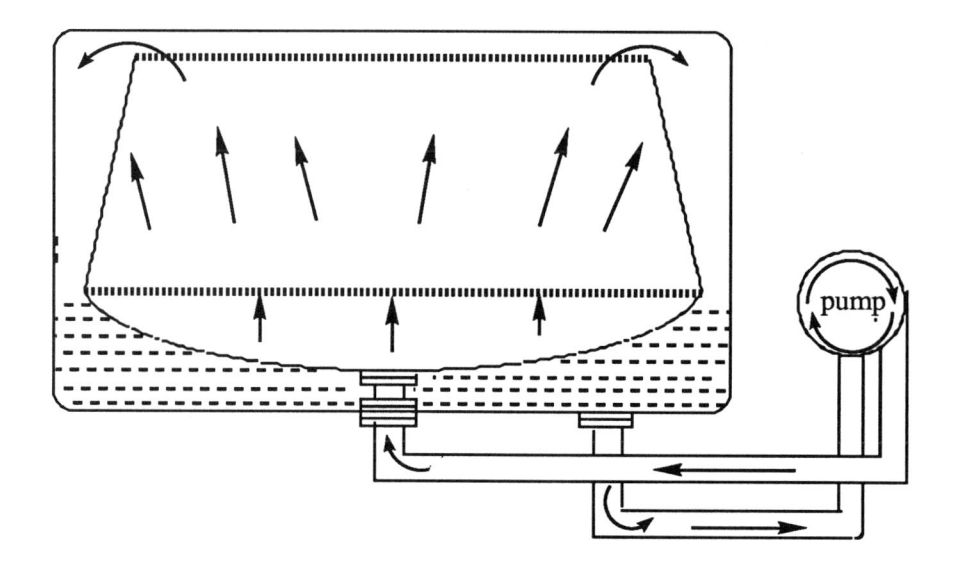

Fig. 5.1 Raw Stock Dyeing Machine

The inner conical pan, in which the fibers are tightly packed, has a perforated base and a perforated top so that the circulated dye-solution can pass through the fibers. The bottom of the conical pan is connected to a pump that forces the dye-solution upwards. The pan is made conical so that when the liquor flows, the fibers are forced upward and pressed against the conical sides, closing channels in the fiber mass and creating a more uniform flow through the fibers. When the solution reaches the top, the dye-solution flows to the bottom of the outer container to which the return side of the pump is connected. The circulation of the dye-solution can be reversed. Also, the entire machine can be sealed so that dyeing can take place under pressure at temperatures above the boil.

5.2 Dyeing Yarns

Yarns can be dyed in the form of skeins or packages. Package-dyed yarns are more suitable for woven fabrics, whereas skein-dyed yarns are more suitable for knits and carpets where a fuller bulk is more desirable. However, package dyeing can be performed on a much larger scale and with more uniform results. Consequently, it is the more commonly-used method.

Package Dyeing Machines

In these enclosed machines (Fig. 5.2) where dyeing under pressure can take place, the dye-liquor is circulated through wound packages of yarn until the dye is evenly exhausted. The yarn is wound onto perforated tubes or springs, or other types of holders, and the packages formed are then mounted onto a perforated rod (spindle) or tube. In a commercial dyeing machine (Fig.5.3) several hundred packages are tightly packed on a number of spindles which are arranged vertically on a hollow base (the carrier). After loading, the carrier is droped into a seating in the dyeing tank, through which the dye-liquor is circulated. The dye-liquor is pumped through the packages in either direction, according to need.

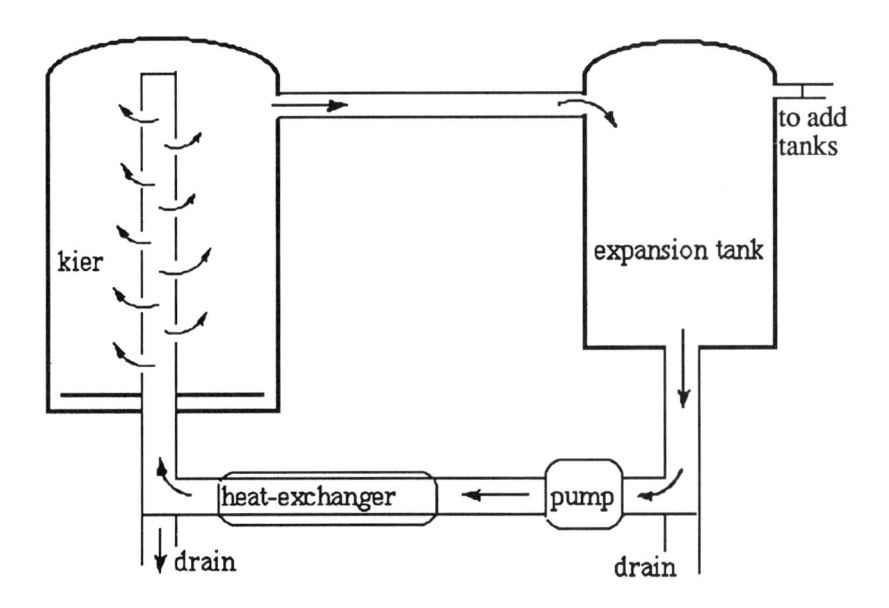

Fig. 5.2 Raw-Stock and Yarn Dyeing Machine

These type of machines are also used, with the proper modification, to dye loose fibers, tops and sliver, and warp yarns. In dyeing warp yarns, a single perforated cylinder of the yarns fits into the seating of the dye tank.

Packages that are too loosely wound may collapse during the dyeing process. However, packages that are too tightly wound, may interfere with the circulation of the dye-liquor. In any case, the packages must be wound as uniformly as possible.

Sudden pressure changes should be avoided to prevent possible distortion of the packages. Therefore, newer package dyeing machines have the capability of controlling both flow and differential pressure.

Many times the packages are covered by a protective bag which acts as a filter to prevent deposits of insoluble dye and other impurities on the yarns.

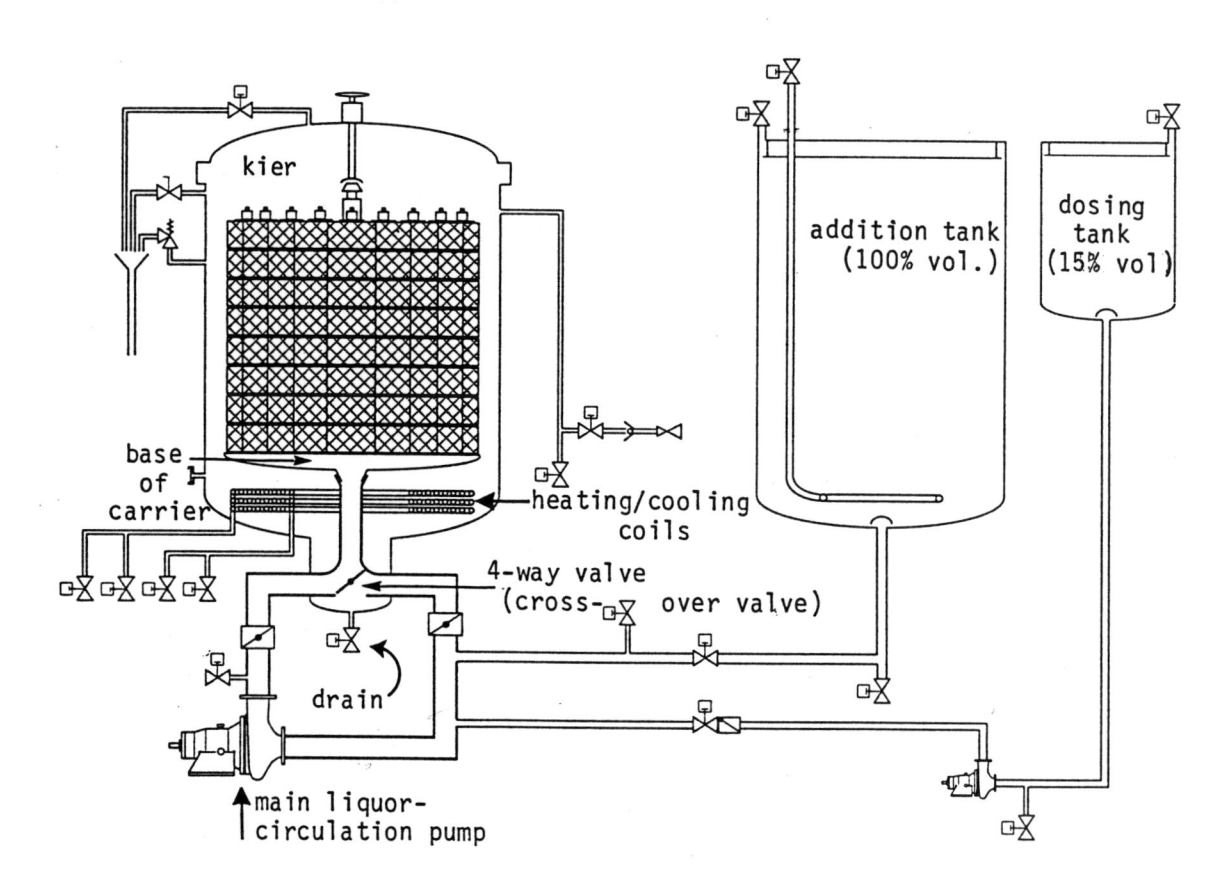

Fig. 5.3 Thies Eco-Bloc Series X Dyeing Machine
Courtesy of Thies GmbH & Thies Corporation

Modern package dyeing machines are made to operate at low liquor ratios. This is achieved by eliminating the external expansion tank. An example of this type of a machine is the 'eco-bloc X' made by Thies GmbH & Co. (see Fig. 5.3). The eco-bloc X operates at liquor ratios of 1:4 to 1:6. The pump runs in one direction, but the liquor flow direction can be reversed by the cross-over valve, situated below the dyeing kier. The cross-over valve directs the flow inside-out or outside-in, and the flow or differential pressure is adjusted by the use of a speed controlled motor. By using the proper carrier for the substrate to be dyed, the eco-bloc X can be used for dyeing packages, warp beams, loose stock, etc.

5.3 Dyeing Fabrics

The Beck

The beck is one of the oldest dyeing machines known. It consists of a tub containing the dye liquor, and an elliptical winch or reel which is located horizontally above the dye-bath (Fig. 5.4).

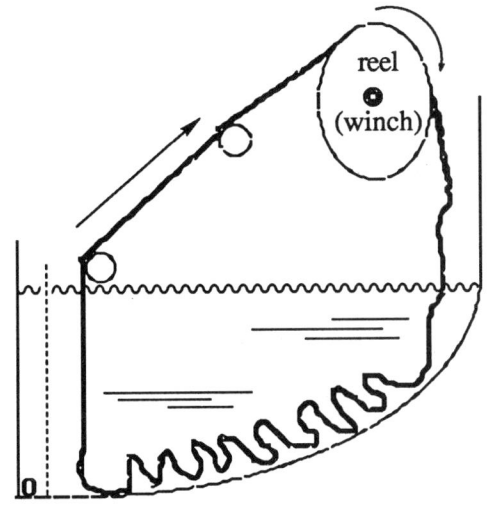

Fig. 5.4 Beck Dyeing Machine

Ten or more pieces of fabric (50 to 100 yards each) are dyed simultaneously. Each piece is drawn over the winch, and its two ends are sewn together to form an endless rope. The ropes are kept in the dyeing machine side by side, separated from each other by rods to prevent them from tangling. During the dyeing process the reel rotates, pulling the ropes out of the dye-bath and dropping them back into the dye-bath at the opposite side. In this way almost all the fabric is kept inside the dye-bath. A vertical perforated partition is located a few inches from the front of the machine, enclosing the steam pipes, so as to prevent them from coming in direct contact with the fabric. This partition also provides a space for adding dyes and dyeing assistants in their concentrated solutions form in such a way that they don't come directly in contact with the fabric.

Becks are used for dyeing knits and other light-weight fabrics that can be easily folded into a rope form without causing damage. Fabrics made of filament yarns that tend to break should not be dyed in a beck since the broken filaments will dye deeper. Very light fabrics should also be avoided as they may tend to float on the dye-bath and tangle.

In becks, dyeing is carried out at high liquor ratios (in the range of 1:20 to 1:50) and since the tension applied by the pool of the winch is low, the dyed fabrics acquire a softer hand. However, the high liquor ratio limits the use of dyes with low substantivity (e.g. reactive dyes). Becks for cotton fabrics provide for dyeing at lower liquor ratios, where as in becks for wool fabrics, higher liquor ratios can be used without loosing substantial amounts of dyes.

One of the main problems with becks is the difficulty in maintaining a uniform temperature throughout the dye-bath. To improve the distribution of heat uniformly, modern becks are enclosed and a pump is added to circulate the dye-liquor.

The Jig Dyeing Machine

In this machine, the fabric is dyed in its open or full width. The machine consists of a relatively small tub and two drawing rollers located above the dye-bath (Fig. 5.5).

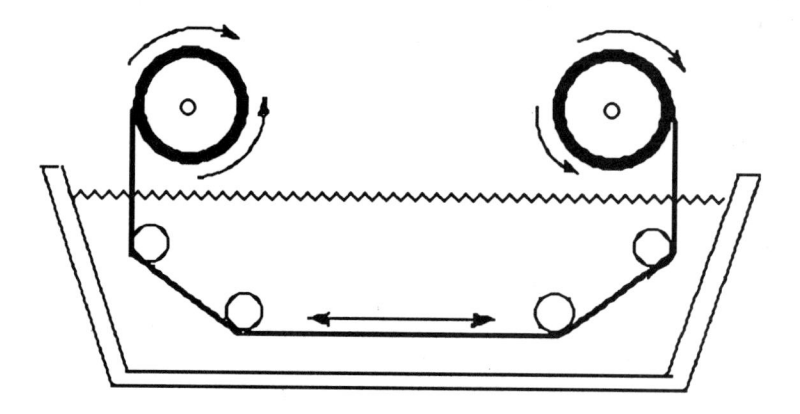

Fig. 5.5 Jig Dyeing Machine

First the fabric is wound around one of the rollers. During dyeing the fabric is passed through the dye-bath as shown in Fig. 5.5, and rewind onto the second roller. When all the fabric is passed through the dye-bath the direction of movement is reversed, and in this way the fabric is passed back and forth throughout the dye-bath (an even number of times to insure uniformity) until the dyeing is completed. Guide rollers are located at the bottom of the tub to insure good contact between the fabric and the dye-bath.

The jig is suitable for dyeing delicate fabrics. Since the dyeing is carried out at low liquor ratios (1:2 to 1:6), the consumption of chemicals and energy is low. In addition, modern jigs are totally enclosed to prevent cooling and loss of heat from the fabric on top of the rollers. Dyeing can be carried out successfully at the boil, but dyeing under pressure is not easy since there are difficulties in operations. The newer jigs are equipped with devices that allow a reduction in the tension applied length-wise. This reduction in tension is essential for delicate fabrics.

A major disadvantage of jig dyeing machines is their low efficiency. During dyeing, almost all of the fabric is outside the dye-bath, and the temperature at the rollers is significantly lower than that of the dye-bath. Also the machine is not suitable for washing so that pre-scouring and/or after-scouring may have to be performed in other machines.

Jet Dyeing Machines

The earliest commercial jet dyeing machine was offered by Gaston County Co. in 1961. In a jet dyeing machine, a very efficient contact between the dye-liquor and the fabric is obtained since both are in constant movement. This results in improved level dyeing and a shorter dyeing time. Jet dyeing machines are similar to becks in that the fabric is circulated through the dye-bath in the rope form. However, in a jet, the transportation of the fabric occurs by circulating the dye-liquor through a venturi jet, instead of the mechanical pull of the reel in a beck. The fabric is pulled out of the main dyeing chamber by means of a high speed flow of dye-liquor that passes through the venturi opening. The venturi tube has a narrowed passage through which the pumped dye-liquor is forced to move much faster. The jet flow of the dye-liquor formed causes a suction that forces the fabric to pass through the jet tube that leads back into the main chamber.

Jet dyeing machines are pressurized and dyeing can take place at temperatures as high as 135°C (275°F). Jets are built to be used at low liquor ratios, between 10:1 and 5:1 and lower. Because of the high speed in which the fabric moves, creases change their position rapidly and there is not enough time to set them in one location. Therefore, creasing is minimal.

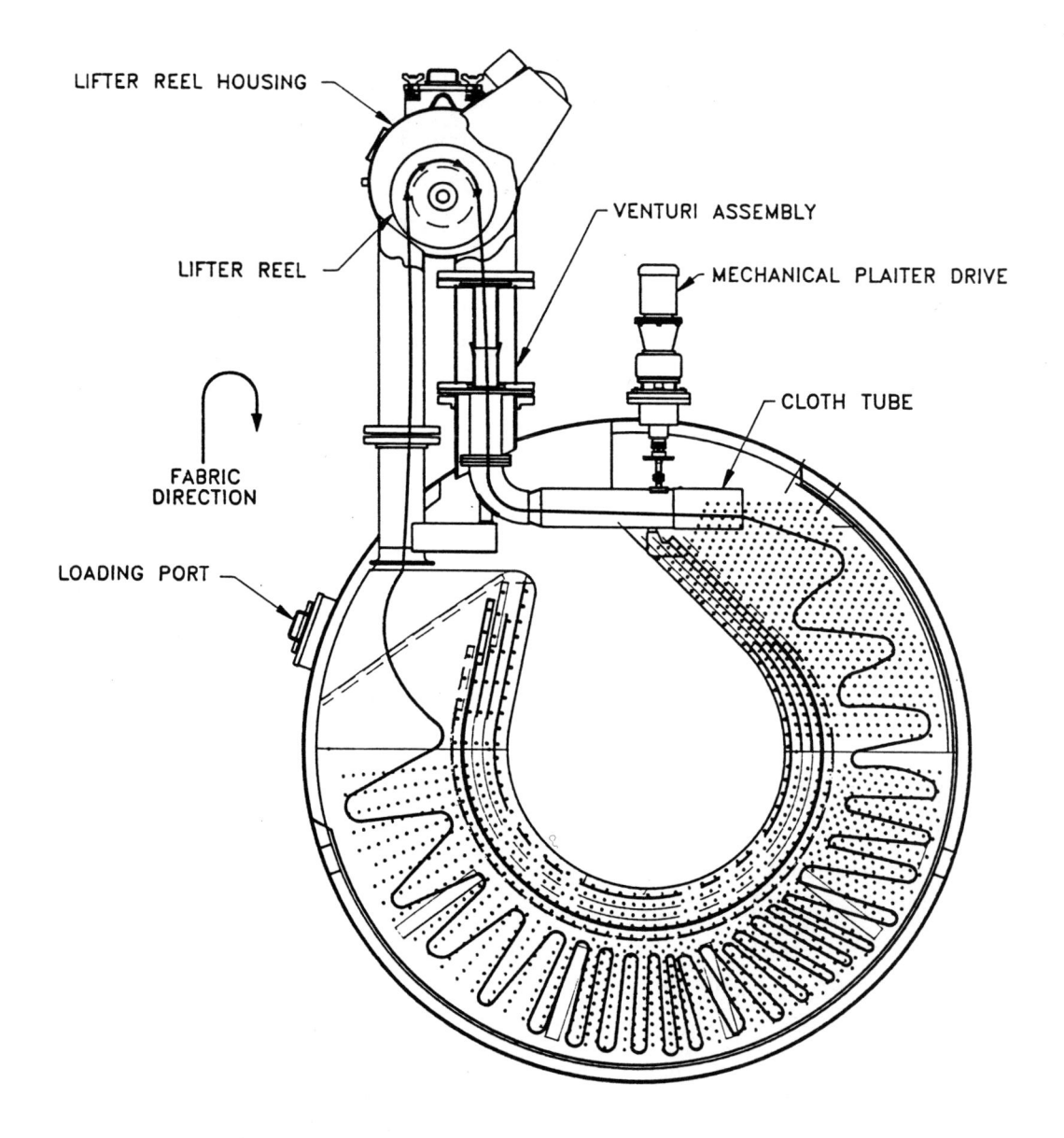

Fig. 5.5 Gaston II 824 Jet Dyeing Machine
Courtesy of Gaston County Co.

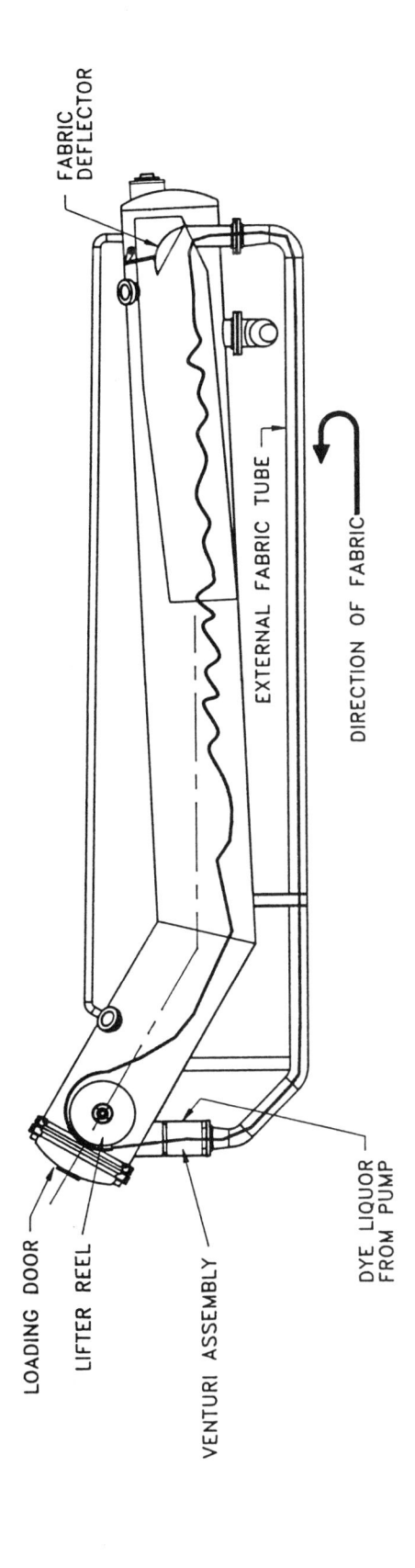

Fig. 5.7 Gaston Futura Jet Dyeing Machine
Courtesy of Gaston County Dyeing Machine Co.

The pull by the venturi jet has been found to be too powerful for certain delicate fabrics and in order to reduce tension, other machines have been developed. In newer jet dyeing machines the fabric is pulled out of the bath by means of a driven lifter reel and then passed through a venturi opening. Examples of these type of machines are the Gaston II 824 jet dyeing machine (Fig. 5.6), and the Futura jet dyeing machine (Fig. 5.7). In these machines, the fabric is lifted first by a driven lifter reel and then it is passed through a venturi jet of a lower velocity than that of a machine that the fabric is driven only by the venturi jet.

Modern jet dyeing machines are generally categorized as 'round kier' or 'cigar kier' configurations. Most fabrics can be dyed satisfactorily in conventional round kier dyeing machines such as the Gaston 824 jet dyeing machine (Fig. 5.6). These type of machines operate at low liquor ratio and yield very good results on most fabrics. However, certain fabrics have more tendency to develop 'crush' or 'pile' marks due to their constructions. This condition is more pronounced in round kier machines due to the compacting pressure on the fabric in the chamber. Dyeing machines with long, horizontal kier designs such as the Gaston Futura (Fig. 5.7), are better suited for these fabrics. The fabric weight is distributed within the long, horizontal ('cigar' shaped) kier which minimizes compaction.

5.4 Equipment for Continuous Dyeing

Successful dyeing in this method is highly dependent on the machinery used. Continuous ranges vary greatly in their components. The most common components found in continuous ranges are: padders, drying and curing ovens, steamers, and washers. An example of a continuous range for dyeing polyester/cotton blends is shown in Fig. 5.10.

The Padder

Padders are used to impregnate fabrics with liquors containing dyes, dyeing assistants or other chemicals. Padding is usually followed continuously by other treatments, from drying to a series of successive treatments.

The simplest padder consist of two parts: the trough containing the dye-liquor, and two squeezing rollers arranged above the dye-liquor (Fig. 5.8). In the padding process, the fabric in its open width form, enters the trough through tension rails, passes through the dye-liquor, and is then squeezed (nipped) between two heavy rubber rollers (nip rollers) with the same proper hardness, under pressure. Excess dye-liquor runs back into the trough. The purpose of the nip is to remove excess dye-solution, and achieve an even application of the dye solution through the full width of the fabric. The high pressure also

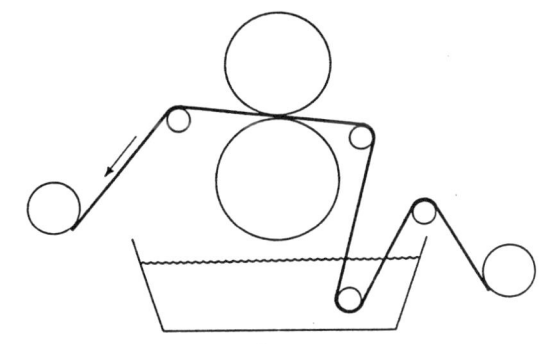

Fig. 5.8 Vertical Padder

forces the dye-liquor further through the fabric, which is particularly important when dyeing tightly woven fabrics. The amount of wet pick-up depends on the pressure applied and the speed of the traveling fabric. A low wet pick-up is desirable when drying follows the padding. This is not only because less energy will be required in the drying step, but mainly to prevent migration of dye with the evaporating water (explained latter).

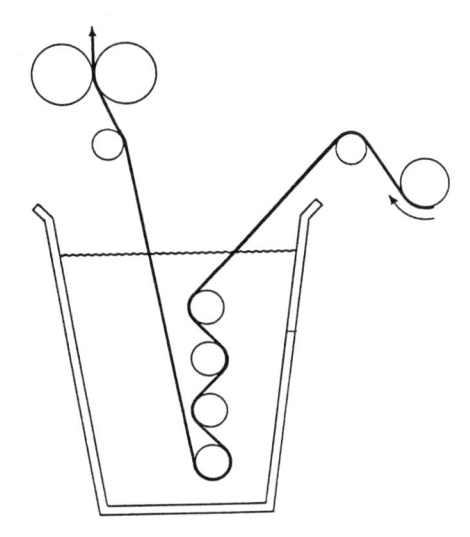

Fig. 5.9 Horizontal Padder

The most common padders are vertical ones with two or three rollers which permit one or two immersions in the dye-liquor (Fig. 5.8) In other types of padders, the squeezing rolls are positioned horizontally (Fig. 5.9). Uneven tension at the padder can cause formation of creases. The pressure across the padding rolls should be controlled to avoid uneven impregnation which may cause differences from selvage to selvage and from selvage to center. The two-roll padders with 'floating rollers' such as those first introduced by Kuester, where the pressure is independently controlled at several locations along the padding rolls, allow the dyer to apply an even pressure and avoid shade differences.

Usually the padder is equipped with a circulating pump and a filter to remove lint and other impurities coming off the fabric into the padding-bath. Lint adhering to fabric during the padding stage may cause specking.

Dye-Selection

Unlike exhaust dyeing, in padding, the dye is applied to the fabric surface all at once. Therefore, the concentration of dye in the impregnating bath is many times higher than the concentration of dye in an exhaust dye-bath. In continuous dyeing, the dye-bath must be free of any undissolved or undispersed dye particles which may cause dye spots. Therefore, dyes with low solubility that are used in exhaust dyeing may not be suitable for continuous dyeing (e.g. certain direct dyes). For the same reason, dyes in liquid form, which are easier to dissolve or disperse, are preferred. The concentrated dye solution or dispersion is filtered or strained prior to its addition to the dye-bath.

One of the problems associated with padding applications is preferential absorption. Ideally, the fabric should pick up the padding liquor as is. When some exhaustion takes place during padding, the fabric will pick up a higher concentration of dye than that of the padding-bath. This will decrease the dye concentration in the padding-bath, and the padded fabric will progressively dye lighter. This shading problem is referred to as 'tailing'. Obviously, dyes with little or no substantivity at the padding stage are preferred. Also, to prevent tailing the trough is made to hold only a small amount of dye-liquor, and fresh dye-liquor is added continuously through a feeding tank. In this way, when preferential dye-absorption takes place, a state of equilibrium will be reached in a short while, after which the shade will remain unchanged.

Drying Equipment

When impregnation is followed by drying, dye-migration becomes a major concern. Evaporating water tends to carry with it dye particles from wet spots to dry spots on the fabric, and from inside or back to face of fabric, and may lead to unlevel and/or shading problems. To prevent migration, drying is done gradually, and preferably, a pre-drying step is included.

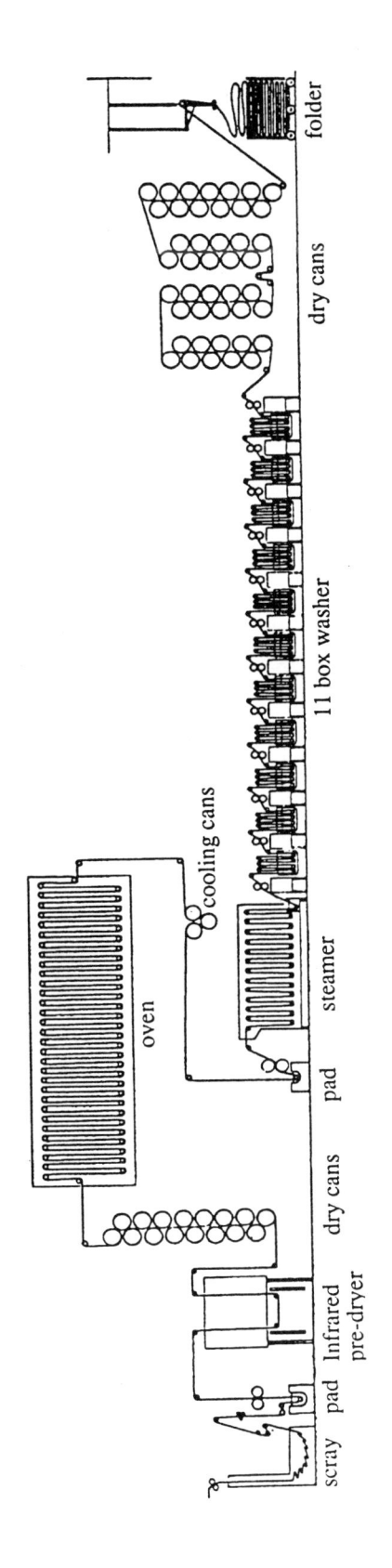

Fig. 5.10 Combined Thermosol-Pad-Steam Continuous Dyeing Range Courtesy of E.I. duPont de Nemours, Inc.

Pre-drying is achieved with infra-red radiation. At this stage approximately half of the water is removed, reducing the wet pick-up to about 30%. The infra-red oven is usually placed directly above the padder so that the padded fabric enters the oven directly without running over guide rollers which may be contaminated with the wet fabric. The infra-red oven itself is a non-contact oven in which the fabric is passed between the infra-red heaters so that radiation is evenly applied to both sides of the fabric. The Infra-red units are heated with gas or electricity, and are mounted vertically one on top of the other with a total height of 2 to 3 yards. In other types of infra-red ovens, the heating units are placed next to each other horizontally, and the fabric is passed over teflon-coated guide rollers located outside the heating units.

The final removal of moisture is carried out over heated cans, or in a hot-flue oven. Heated 'cans' are steam-heated cylinders. Steam under pressure is circulated through the inside of the cans, and the fabric is passed over their surface which is about 2yd. in circumference. The cans are mounted in banks vertically so that the passing fabric comes in contact with the maximum area for each can (see Fig. 5.10). A typical dryer contains two to three dozen cans with controls to adjust their temperature individually. In order to carry out the final drying carefully, the temperatures of the cans are gradually increased. For example, the first group of cans is set at temperatures in the range of 71°C (160°F), and the last ones are set at 130°C (266°F) or higher, according to need.

In a hot-flue dryer, the fabric is passed through the oven over a set of parallel upper and lower guiding rollers, in a winding path. The upper motor-driven rollers pull the fabric through. The lower rollers rotate freely, only guiding the fabric along its path. Air is circulated along heat exchangers, located at the bottom of the oven, up toward the fabric.

Fixation Ovens

These ovens are used when fixation of the dyes is performed with dry heat. Both hot-flue or heated cans are used for this purpose. Since temperatures as high as 215°C (420°F) are often required, the cans are heated with hot oil or gas. Contact heating, as with heated cans, has the advantage that less time is required for the fixation process as compared to the use of dry air.

Steamers

Steamers are used when fixation is carried out in presence of steam. Steam is applied in different forms according to need. Steam can be used in the following forms:

1. Wet. The atmosphere in the steamer is saturated with water in the form of small droplets. The temperature in the steamer is at 100°F (212°F) or a few degrees higher. Under these conditions the drying of the fabric will not occur during steaming.

2. Dry steam. The steamer is saturated with water in the vapor phase only.

3. Super-heated steam. Water vapors are at atmospheric pressure and a temperature higher than that needed for complete evaporation (i.e. above 100°C).

Wash-boxes

Eight or more wash-boxes complete the dyeing process. In the wash-boxes, chemical reactions, as well as washing, take place. The wash-boxes are used to carry out rinsing, neutralization, oxidation, soap at the boil, and other after-treatments, all according to need.

References

1. Engineering in Textile Coloration, Edited by C. Duckworth, 1983, The Dyers Company Publications Trust, West Yorkshire BD1 2JB, England
2. Textile Dyeing Operations, S.V.Kulkarni et al, 1986, Noyes Publications, Park Ridge, new Jersey

Chapter 6.

PREPARATORY OPERATIONS

The main objectives of preparation treatments of textile materials are:

a. remove from the fibers all impurities, both natural and/or those added during production, that may interfere in subsequent dyeing and/or finishing applications,

b. improve the ability of the fibers to absorb water solutions of dyes and chemicals,

c. impart the proper brightness or whiteness to fibers according to need, especially when brilliant or certain pastel shades are desired, and,

d. impart dimensional stability to thermoplastic textile materials.

The importance of adequate and uniform preparation prior to dyeing cannot be overemphasized. Improper removal of impurities can lead to unlevel dyeing, streakiness, and poor penetration. It is estimated that more than 60% of faulty dyeing are the result of improper preparation.

As mentioned above, preparation procedures may vary greatly from one fiber type to another. While natural fibers usually require extensive scouring and bleaching, synthetic fibers may need only a mild scouring before they are ready to be dyed. In cases where removal of sizing and other impurities is very difficult to accomplish, the cleaning procedure may also include scouring with a solvent.

The type of equipment used to carry out the various preparation applications varies greatly from one preparation range to the other. Equipment for batch applications (e.g. Kiers, washers, and dyeing machines), as well as for continuous applications are used. Fabrics are treated in continuous equipment both in the open width or in the rope form.

The following is a general procedure for cleaning fibers, yarns, or fabrics, with small amounts of impurities:

The scouring bath is prepared with 0.1-0.2% of a detergent, and a similar amount of a base such as sodium carbonate or tri-sodium phosphate. Usually a mixture of an anionic and a nonionic detergent is used, but some prefer to use only an anionic detergent. The scouring machine is run at 60°C to 80°C (140°F to 176°F) (for protein fibers the temperature must not exceed 50°C) for 15-20 minutes. The material is then rinsed first with water at a temperature as close as possible to the scouring bath temperature, and then with cold water.

When only small amounts of impurities are present, dyeing dark shades can sometimes proceed without previous scouring. Adding a small quantity of a detergent to the dye-bath may be sufficient to remove enough of these impurities during the dyeing process.

6.1 Cleaning of Wool

Raw wool contains the following impurities: wool-grease, soil, dust, cellulosic fragments, and suint (dried perspiration, mostly soluble potassium salts). These impurities may represent as much as one to three quarters of the weight of the unscoured wool. Obviously raw wool has to be scoured first.

Protein fibers are sensitive to bases. Scouring procedures for cotton may include a base treatment with 3% Na_2CO_3 at the boil. However, when this treatment is applied to wool it will seriously damage it. Wool can be scoured with a detergent solution at a pH of up to 11 (e.g. 0.4% Na_2CO_3), when the scouring-bath temperature is kept below 125°F

In a typical scouring procedure of raw wool the fibers are treated at temperatures below 60°C (140°F), with an anionic or a non-ionic detergent and a small amount of a weak base such as sodium bicarbonate or ammonia. Though not all of the impurities are removed by this treatment, this cleaning is sufficient for preparing the wool for spinning. The wool, however, will be scoured again prior to dyeing, and the complete removal of cellulosic material will be achieved by 'carbonizing'.

The carbonizing process clearly demonstrates the difference between the behavior of wool and cellulose under acidic conditions. In carbonizing, cellulosic materials such as leaves, grass, seeds, and vegetable fiber, are removed from wool. The fabric (sometimes loose fibers) is treated with a diluted solution of H_2SO_4, then dryed, and cured. Under these conditions, the cellulosic materials decompose to form a very fine black powder which can be easily dusted out, while the wool fibers remain unharmed.

6.2 Bleaching

Bleaching is the process of removing or decomposing colored impurities that cannot be removed from fibers by extensive scouring. It is performed on all-white fabrics to produce a high degree of whiteness. Bleaching is also required for fibers that are to be dyed in pastel shades, in particular light blues and violets, and for materials to be dyed in colors with maximum brightness.

In the process of bleaching, colored materials are oxidized and converted into colorless substances. The oxidation reaction, however, is not a selective one, and the fibers themselves may undergo oxidation as well. Improper use of the bleaching agent or excessive bleaching of a cotton fabric, for example, may not only cause tendering but may even lead to the formation of holes in the fabric. The main objective of the bleaching process is, to achieve the desirable level of whiteness with minimum damage to the fibers, and within the shortest possible time.

Currently, hydrogen-peroxide is by far the most widely used bleaching agent. Sodium chlorite (Na ClO_2) and sodium hypochlorite (NaClO) are also being used as bleaching agents but not frequently.

Bleaching is applied more often to natural fibers, in order to remove their natural pigments, and other colored impurities such as plant fragments. The synthetic fibers, however, are usually synthesized in a clean form and appear white. Nevertheless they may have to be bleached if their level of whiteness is unsatisfactory. Also, discoloration of synthetic fibers may take place during their processing. This could be, for example, the result of yellowing that may occur during heat-setting or certain chemical treatments.

Cotton can be bleached with any of the three bleaching agents mentioned above. Sodium hypochlorite has long been used as a bleaching agent for cotton. However, it has been gradually replaced with hydrogen peroxide which is less expensive and can be used successfully in continuous methods and under various conditions. Hydrogen-peroxide is the bleaching agent of choice for wool. As to the synthetic fibers, in particular acrylic fibers, and to a lesser extent nylon and polyester, they are bleached effectively with sodium chlorite, where as hydrogen peroxide and sodium hypochlorite may not be adequate for discoloration of certain synthetic fibers.

6.3 Bleaching with Hydrogen Peroxide

The advantages of using hydrogen peroxide as a bleaching agent are numerous. It can be applied to various fibers, and under a wide range of application conditions. Bleaching with hydrogen peroxide can be carried out by batch or continuous methods, and at room or high temperatures. In addition, hydrogen peroxide is relatively inexpensive, does not release toxic chemicals or unpleasant odors, and does not cause corrosion of equipment. The only by products released upon its complete decomposition are water and molecular oxygen. When using hydrogen peroxide on fibers that are sensitive to oxidation, such as wool or cotton, damage can be kept to a minimum provided that the bleaching is carried out carefully under the recommended conditions (pH, temp., etc.).

Hydrogen peroxide is applied under basic conditions, and the optimum pH for bleaching is between 10 and 11. At a pH below 10 hydrogen peroxide is quite stable, and oxidation will occur too slowly. When the pH of the bleaching bath is above 11, hydrogen peroxide decomposes too fast, and may cause damage to the fibers.

The mechanism by which hydrogen peroxide acts as an oxidizing agent is believed to take place through its perhydroxy ion (HO_2^-). When dissolved in water, hydrogen peroxide behaves as a very weak acid and dissociates according to:

$$H_2O_2 \rightleftharpoons H^+ + HO_2^- \quad \text{(dissociation constant } 10^{-12}\text{)}$$

The relatively unstable HO_2^- anion decomposes slowly to yield monoatomic oxygen:

$$HO_2^- \longrightarrow OH^- + O$$

Bleaching, most probably takes place via the reaction of HO_2^- or O with colored materials where by they are oxidized into colorless compounds. In absence of other chemicals monoatomic oxygens combine to yield molecular oxygen and the oxidation power of hydrogen peroxide is lost.

Since the amount of HO_2^- at pH 7 is extremely small, neutral solutions of hydrogen peroxide are quite stable. Commercial solutions of hydrogen peroxide contain small amounts of stabilizers that prevent their decomposition while in storage.

In order to activate hydrogen peroxide for the bleaching reaction to take place, a base is added to the bleaching bath. The base will shift the equilibrium dissociation to the right:

$$NaOH + H_2O_2 \rightleftharpoons Na^+ + HO_2^- + H_2O$$

Thus, increasing the pH of the bleaching bath will increase the concentration of HO_2^- ions, and so will increase the rate of oxidation.

Besides hydrogen peroxide and a base (sodium hydroxide or sodium carbonate), the bleaching bath contains stabilizing agents. Sodium silicate - $Na_2SiO_3 \cdot 5H_2O$ has been found to be the most suitable stabilizer. It is a base that acts as a buffer at the proper pH of 10.5-11.5, and in addition, it prevents the catalytic decomposition of hydrogen peroxide caused by heavy metals. The catalytic effect of transition metal ions (e.g. Fe^{+++} or Cu^{++}) on hydrogen peroxide in accelerating its decomposition may cause great damage to the fibers. Sodium silicate also acts as a sequestering agent and forms with transition metals soluble complexes. Mg^{++} ions, however, have been found to posses a stabilizing effect on hydrogen peroxide. Accordingly, when using soft water a small amount of a magnesium salt is added to the bleaching bath.

Note that sodium silicate is not a powerful sequestering agent. It forms complexes with transition metals but not with alkali earth metals such as Mg^{++} and Ca^{++}.

A major drawback in using sodium silicate is that it can form insoluble deposits on the goods as well as on the dyeing equipment. These deposits which are extremely difficult to remove, give the fabric a harsh handle, and may lead to unlevel dyeing. The problem may occur when a final rinse with an acid is given, after the bleaching, to neutralize the fabric. Any residual sodium silicate left on the fabric will change into insoluble salicilic acids. Consequently other stabilizers, inorganic and organic, have been developed to replace or reduce the amount of sodium silicate in the bleaching bath.

Another source of hydrogen peroxide is the solid sodium percarbonate: $2Na_2CO_3.3H_2O_2$, which when dissolved in water dissociates into sodium carbonate and hydrogen peroxide. A 1% solution of sodium percarbonate has a pH of 10.5 and contains the proper amount of hydrogen peroxide for bleaching. The main disadvantage of this chemical, is that when exposed to moisture in storage, it will slowly decompose, thus releasing the oxygen prematurely.

6.4 Preparation of Cotton Fabrics

Cotton fibers on the average consist of 90-96% cellulose when bone-dry. The approximate amounts of other materials present in the fiber are: 1.1-1.9% proteins, 0.7-1.2% pectins, 0.4-1.0% waxes, 0.7-1.6% ash, and 0.5-1.0% other impurities. These materials are located mainly in the primary walls of the fiber, that is at or near the fiber's surface. Because of their hydrophobic nature it is difficult to wet-out unsecured cotton. In the scouring process these impurities are removed to the extent that the fibers will become hydrophilic and will easily wet-out.

In addition to the natural impurities, cotton fibers contain other foreign materials. Raw cotton is mechanically cleaned at the gin. Additional mechanical cleaning takes place at the first stages of manufacturing of the yarns. Nevertheless, a considerable amount of fine particles of broken seeds and other fragments of the cotton plant are still left attached to the fibers. These impurities, technically known as motes, appear on unbleached yarns or fabrics as dark specks.

The natural waxes of cotton serve as protecting and lubricating agents during yarn manufacturing and during weaving or knitting. Therefore, raw cotton is normally not scoured until dyes or finishing chemicals need to be applied.

During the processing of fibers into yarns and during weaving or knitting operations, additional impurities are accumulated. Spinning oils are used in knitting, and sizing materials are applied to the warp so that it won't break easily during weaving. Woven fabrics coming out of the loom may contain as much as 20% foreign substances, natural and added materials, most of which has to be removed during the preparation of the fabric, prior to dyeing.

The preparation of fabrics made of cotton or polyester/cotton blends may include some or all of the following operations: singeing, desizing, scouring, bleaching, and mercerizing. With increasing demands for savings in chemicals, energy, and water, certain treatments are often combined. An example of a combined application is the so called 'oxidative desizing' where desizing and scouring are taking place at the same time.

Singeing

Before the removal of impurities by wet-processing begins, fabrics are usually singed. In this process fuzzy and protruding fiber ends are removed by burning them off. Singeing is done to fabrics that require a clear and smooth surface such as broad-cloth and batiste, and fabrics to be printed, and it is usually performed on both sides of the fabric.

In a singer, a fabric is passed over open flame at very high rates (e.g. 300 yards per min.) to prevent scorching, after which the singed fabric is passed through water (or introduced directly into the desizing bath) to extinguish any sparks that may cause it to ignite.

Uneven singeing may lead to unlevel dyeing. Therefore, the fabric is passed through the singer flat, in open width and under a slight tension. The gas burners should be properly controlled and maintained; otherwise streaky dyeing may occur. Also, the same level of singeing must be applied to face and back of fabric, to avoid face to back shading.

The singeing of a fabric containing thermoplastic fibers such as polyester will also reduce its tendency to pill. However, the heat applied to the fabric may cause protruding fibers to melt and form beads. These beads in turn tend to dye darker. Also, there is a possibility of fixing various impurities and sizing materials in thermoplastic fibers. Therefore, it is better to singe certain blends of synthetic fibers with cotton or wool after scouring, and sometimes even after dyeing.

Desizing

Prior to weaving, cotton warp yarns are coated with sizing materials to provide them with additional strength, and resistance to abrasion. The most common sizing materials used are different kinds of starches, and various grades of partially hydrolyzed polyvinyl alcohol. Among the other sizes used are soluble cellulose derivatives, and soluble polyacrylates. A typical formulated sizing mix often used, is a mixture of a starch, partially hydrolyzed polyvinyl alcohol, and a small amount of a lubricating agent, such as a fat or a paraffin wax, to assist in the smoothness of the weaving operation. The increase in weight obtained due to sizing is in the range of 20% (on the weight of the warp yarns).

Water soluble sizes such as partially hydrolyzed polyvinyl alcohol, and cellulose derivatives, are removed with hot water. More difficult to remove are the insoluble degradation products.

Enzymatic Desizing

Enzymes are often used for the effective removal of starches. The enzymes are complex proteins found in vegetable and animal cells. They act as highly specific catalysts in chemical reactions that take place within living cells. There are enzymes that catalyze the hydrolysis of starches (e.g. an alpha amylasa type). In their presence, hydrolysis of starch takes place under mild conditions, to yield soluble dextrins and glucose oligomers. Though both starch and cellulose are polymers of glucose, the catalytic effect of the enzymes used is specific to the hydrolysis of starch. These enzymes have no effect at all on cellulose so that the cotton remains unchanged during the enzymatic desizing.

Enzymes are sensitive to their environment. They have a limited shelf life of only several months, and should be stored at temperatures below 21°C (70°F). When used at temperatures or pH other than those recommended they will be destroyed.

Most desizing enzymes have their optimum temperature operation in the range of 38-65°C (100-150°F). There are, however, enzymes that operate well at the boil and above. Within the recommended temperature range of a specific enzyme, there is a temperature where the enzyme shows maximum activity. The pH of the impregnating bath, which is usually in the range of pH 4.5-7.5, has to be adjusted properly so that the enzyme will perform with high efficiency.

In a typical application the fabric is passed slowly through a saturator where it is thoroughly impregnated with the desizing bath and then stored for 8 to 12 hours usually in rollers. The impregnating bath is prepared with the required amount of the enzyme, a wetting agent, and a proper salt (e.g. sodium polyphosphate or NaCl, according to the recommended formulation). A detergent may also be included provided that it will not deactivate the enzyme as some detergents (or wetting agents) may do. After batching, the fabric is thoroughly washed with hot water.

When the enzyme is applied by padding, care should be taken to insure a sufficient pick up of the padding bath (80-90% w.p.u. for 100% cotton, and 70-80% w.p.u. for 50/50 polyester/cotton). At this stage the fabric is difficult to wet because of its hydrophobic contents (natural impurities and those added during manufacturing). A wetting agent, and a long dwelling time in the impregnating bath will help in obtaining the required wet pick-up.

Other enzyme application methods are performed at higher temperatures where the desizing can be achieved in a short time as in the continuous pad-steam procedure. In this method the fabric is padded with the enzyme bath and then passed into a wet steam chamber (at 96°C-100°C) (205-212°F) where the desizing takes place in less then a minute.

The main advantage of desizing with enzymes is that there is no risk of damaging the fibers. However, enzymatic desizing is relatively expensive since it cannot be combined with other preparatory operations.

Caustic Scouring

This cleaning treatment, also referred to as alkali boil-off, removes most of the cotton's impurities. This includes natural and other impurities as well as sizing residues that were not removed during desizing. In addition, caustic scouring swells the motes and opens them up. This prepares the motes to react more readily with the bleaching agent in the next step. Therefore, when caustic scouring is applied as a single step it is always performed prior to bleaching.

In caustic scouring the fabric is treated with a strong alkali solution (e.g. 5-10g./l sodium hydroxide, or a mixture of sodium hydroxide and sodium carbonate of a similar alkalinity strength) close to or above the boil, for 1-2 hours. The alkali treatment is followed by a hot rinse to remove the emulsified impurities. The final rinse may include a small amount of acetic acid if the fabric has to be neutralized.

In the course of caustic scouring, natural impurities become water soluble. Pectins are converted into their sodium salts which are water soluble, and proteins undergo basic hydrolysis to form water soluble amino acids. The hydrolysis of the waxes yields soap. The soap thus formed serves as the detergent in the scouring process to emulsify the hydrophobic materials that are still present on the fibers.

A sequestering agent is usually added to the scouring bath to prevent Ca^{++} and Mg^{++} ions from combining with soap molecules. A wetting agent and/or a detergent may also be added to the scouring bath.

Bleaching

Currently almost all cotton and polyester/cotton fabrics are bleached with hydrogen peroxide, when needed. Hydrogen peroxide bleaching is carried out by exhaust, semi-continuous, and continuous methods. Of these, the preferred and most widely used is the continuous bleaching in the open-width. In particular, the open-width method is preferred when bleaching blends of cotton with heat-sensitive fibers, such as polyester, in order to avoid crease marks and other defects caused by treating fabrics in the rope form.

In typical batch treatments of cotton fabrics with hydrogen peroxide in kiers, the bleaching bath is prepared as follows:

hydrogen peroxide (35%)	4-8% (o.w.f.)
sodium hydroxide	0.5-1% (o.w.f.)
sodium silicate	2-4% (o.w.f.)
wetting agent or detergent	when needed

The bleaching is then carried out near the boil or at temperatures above the boil, under pressure, for an hour or more. After bleaching, the goods are thoroughly rinsed with a slightly basic solution to avoid the formation of insoluble silicates.

In a typical continuous bleaching in the open-width the padding bath contains the following:

hydrogen peroxide	(50%) 1.5-2.5%
sodium hydroxide	0.2-0.6%
sodium silicate	1-1.5%
wetting agent	0.1-0.2%

After padding the fabric is passed through a steamer. Steaming time may vary from as little as a few minutes to one hour or more, depending on the type of steam and steamer used.

Combined Operations

These are very attractive options since savings in time, energy, labor, etc., are possible. However, combined operations are not always sufficient for preparation of certain cotton fabrics for dyeing and finishing.

In oxidative desizing, where scouring takes place as well, the cotton fabric is padded with a solution containing sodium hydroxide, hydrogen peroxide, a stabilizer, and a wetting agent. After padding, the fabric is steamed for the proper amount of time, depending on the type of steamer and the concentration of chemicals used. After steaming, the fabric is rinsed with hot water several times to remove the polyvinyl alcohol and other emulsified materials. During the treatment the starch is oxidized and becomes water soluble. Not all the starch is removed in this treatment, and what is left in the fabric will be removed at the bleaching step that follows. Caustic scouring also takes place due to the presence of sodium hydroxide in the proper concentration.

Attempts are made nowadays to use a single stage operation for selected fabrics as described, for example, by the Committee of Palmetto Section of the AATCC [4]. The procedure is similar to that of oxidative desizing, and the degree of whiteness depends on the amount of hydrogen peroxide used. According to this study [4] the minimum levels for effective preparation were:

hydrogen peroxide (50%)   50g./l
sodium hydroxide (50%)   40g./l

Fabric samples were impregnated with the above solution (w.p.u. 100%), and steamed for 20min. at 100°C (212°F). The preparation was completed by rinsing with hot water (82°C-93°C) (180-200°F) for at least 20 min.

Mercerization

The treatment of cotton fabrics or yarns with a cold concentrated solution of sodium hydroxide solution for one minute or less is called mercerization. This treatment is unique since several significant improvements in the properties of the cotton fibers occur in a single step. This process was named after John Mercer who was the first to discover in 1844, that the treatment of cotton with a concentrated sodium hydroxide solution brought about permanent changes in the fiber's properties. In his patent issued in 1850, he claimed that the treatment resulted in significant increase in the tensile strength, water absorbency, and dyeability of the fibers. He also noted that the treated material underwent considerable shrinking: with yarns contracting in length, and fabrics shrinking in both the warp and filling directions.

The excessive shrinkage, resulting in loss of goods, prevented this caustic treatment from becoming commercialized. It took more than forty years until it was found (H. A. Lowe, English patents in 1889 and 1900) that shrinking could be prevented if the yarns or fabric were kept under tension during mercerization. When mercerizing under tension, the fibers also acquire a silk-like luster.

Mercerized cotton can be recognized when examined under a microscope. Cotton fibers are naturally flat and twisted (ribbon-like). During mercerization the fibers swell, untwist, and their oval cross-section changes into a round form.

Besides imparting a greater affinity to water, dyes, and chemical finishes, mercerization also helps to cover differences in dye uptake between regular cotton and 'dead' or immature cotton. Dead cotton is the reason for white or light specks that are found sometimes on dyed cotton.

For the mercerization to produce improved dyeing results the process has to be carefully controlled. Therefore, factors affecting the degree of mercerization must be carefully controlled; otherwise the uneven application will lead to unlevel dyeing. In particular, the amount of tension applied, and the concentration and temperature of the sodium hydroxide-bath should be the same through out the entire application.

A typical continuous mercerization of a cotton containing fabric in the open width is carried out as follows:

The fabric is padded with about 20-25% sodium hydroxide solution containing a wetting agent, and then passed over several cans to allow a dwelling time of approximately one minute during which the caustic solution will penetrate the fibers and react with them properly. At this stage the tension is applied only in the length direction. The fabric is then placed on a tenter frame (tension now is applied in both the warp and the filling directions) and is pulled to its original or desired dimensions. While on the tenter frame, the fabric is washed by spraying water until the amount of sodium hydroxide on the fabric is reduced to only a few percent. Now, the fabric is removed from the tenter frame, and the remaining alkali is removed by passing the fabric through several washers, one of which contains a diluted solution of sulfuric or acetic acid.

Currently, mercerized cotton is produced mainly for its increased dyeability and luster. The need for these qualities is evident when trying to obtain deep shades with the bright reactive dyes. Building heavy shades with certain reactive dyes is difficult unless the cotton has been mercerized before. Also the acquired luster will provide a lustrous background for the dyes that will further enhance their brightness.

6.5 Heat-Setting of Thermoplastic Fabrics

Heat-setting or thermosetting is a heat treatment applied to fabrics made of thermoplastic fibers such as polyester or nylon to impart dimensional stability. Heat-setting temperatures are well above the glass transition temperature ($T^o g$) of the fiber, and heat is applied by means of hot air, heated cans, or steam. The treated fabric acquires dimensional stability, e.i. a memory to the shape it was during the heat-setting. It will resist shrinking and creasing, and will have the ability to maintain pleats present in the garment during the Heat-setting treatment. In a typical heat-setting of polyester, the fabric is placed on a tenter frame set to the desired final width, and introduced into a hot-air oven at a constant temperature chosen in the range of 190-215°C (374-420°F) for 30 to 90 seconds.

A fiber is said to be stable when it is completely relaxed. Stretching or twisting introduces tension into the fiber. Fibers of a fabric coming out of a weaving or knitting machine are under stress. Tensions are introduced into synthetic fibers during spinning and subsequent drawing. Additional tensions are introduced when making the yarns and in weaving or knitting, during which time the yarns are bent or stretched. When the fiber's temperature is raised in heat-setting to well above its $T^o g$, vibrations and other molecular movements increase, weak secondary bonds in its amorphous regions break, and strain is released. This in turn will allow the molecules to rearrange into a new more relaxed and stable configuration, and form new second order bonds. The new stable form will retain its dimensional stability as long as the fabric is not exposed again to similar high temperatures. As a general rule fabrics or garments are heat-set at temperatures that are at least 30°C higher than the highest temperatures that the fabric will be exposed to in subsequent treatments and/or in normal use.

Heat-setting affects the dyeability of the fiber. Usually it decreases its dyeability, and therefore when performed before dyeing it is extremely important to apply the heat-setting uniformly. Uneven temperatures in the oven may cause differences in the fabric from selvage to selvage and/or from selvage to center which will show later as unlevel dyeing.

References

1. W. Prager, BASF, Metting "The Challenge of the 80's" in Pretreatment and Bleaching, The Dyer's World-1980's Theory to Practice, A.A.T.C.C., No. 26, 1980
2. Pretreatment of Cotton Fabrics, Hoechst, January 1983, Frankfurt am Main 80, Germany
3. Smith Brent and Rucker James, Troubleshooting in Preparation- A Systematic Approach, A.D.R. September 1987, p. 34
4. Single Stage Preparation-A Viable Alternative for Selected Fabrics, Research Committee of Palmetto Section, AATCC, Book of Papers, 989, AATCC International Conference & exhibition, p. 43

Chapter 7.

DYEING WOOL WITH ACID DYES

7.1 The Structure of Wool

The micro and macro structures of animal hair are far more complex than any of the other natural and synthetic fibers. Wool is made of the fibrous protein keratin, the structural material of all animal hair. Proteins are polymers composed of different alpha-aminoacids, bonded to each other through amide linkages:

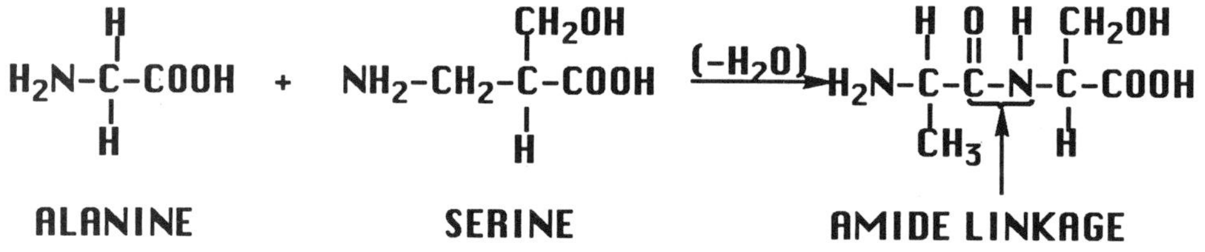

$$H_2N-\underset{H}{\overset{H}{C}}-COOH \quad + \quad NH_2-CH_2-\underset{H}{\overset{CH_2OH}{C}}-COOH \quad \xrightarrow{(-H_2O)} \quad H_2N-\underset{CH_3}{\overset{H}{C}}-\underset{}{\overset{O}{C}}-\underset{}{\overset{H}{N}}-\underset{H}{\overset{CH_2OH}{C}}-COOH$$

ALANINE   SERINE   AMIDE LINKAGE

Fig. 7.1 Formation of an amide linkage

The resulting polymer is a very long polyamide chain of the general formula:

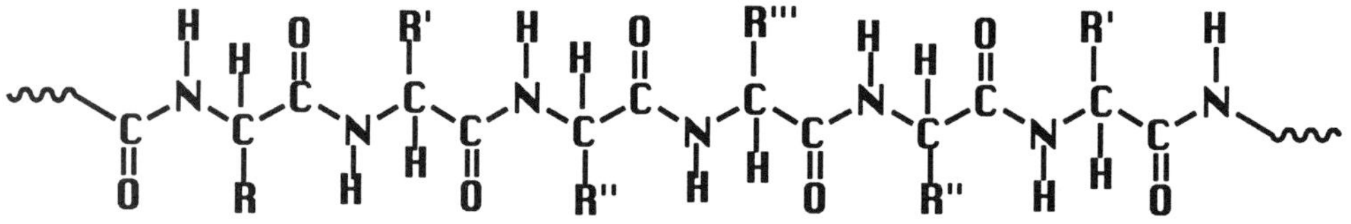

(R, R', R'', ... = -H, -CH$_3$, -CH$_2$OH, -CH$_2$-C$_6$H$_5$, -CH$_2$SH, -CH$_2$-S-S-CH$_2$-, -CH$_2$-CH$_2$-CH$_2$-CH$_2$-NH$_2$, etc.).

Fig. 7.2 A polyamide

About 20 amino acids are found in wool, each with its characteristic short side chain (the functional R group).

Wool contains a considerable amount of the double amino acid cystine which provides disulfide (-S-S-) cross-links within the polyamide chains, as well as between neighboring polyamide chains:

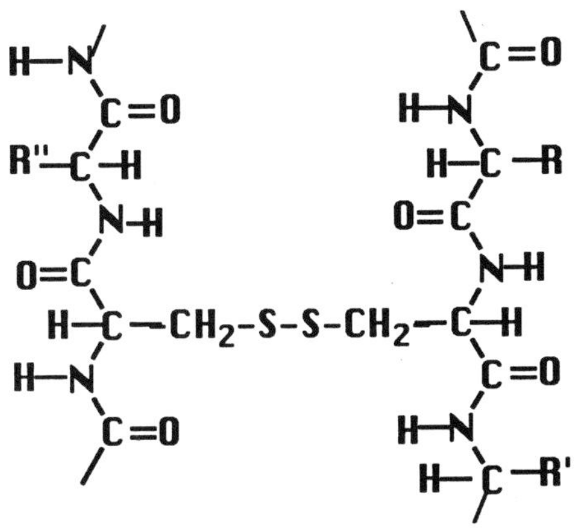

Fig. 7.3 A disulfide cross-link in wool

The strength and stability of wool depends to a large extent on these cystine linkages. Strong alkaline solutions, particularly at high temperatures, damage wool by breaking the cystine cross-links. Setting wool fabrics, as in crabbing (treatment of wool fabric under tension with boiling water for 30 min.) causes some cystine linkages to break, and new cross-links will be formed. Accordingly, when dyeing wool fabrics in the rope form at the boil, they should be treated carefully to avoid tension which could lead to permanent creases.

The Micro Structure of Wool
Several levels of structure are involved in the formation of wool fibers. The exact sequence of the amino acids within the polyamide chain and the location of the disulfide cross-links are referred to as the 'primary structure' of the protein, where as the three dimensional arrangement of the polyamide chain is referred to as the 'secondary structure'. In wool, the long polyamide chains are coiled in a spiral which have the 'alpha helix' configuration. This shape is stabilized by hydrogen bonds formed between the NH and CO groups of the polyamide chain as shown in Figure 7.4 :

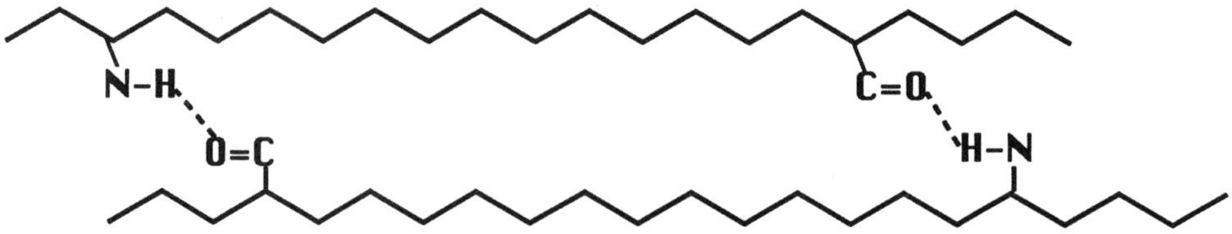

Fig. 7.4 Hydrogen bonds in wool

The presence of ionic cross-links (Fig. 7.5) also significantly contribute to the stabilization of the three dimensional structure of wool. These bonds are formed when a carboxylic group and an amino group from two different side chains, along the same polyamide chain or from two adjacent polyamide chains, are sufficiently close to each other (Fig. 7.5):

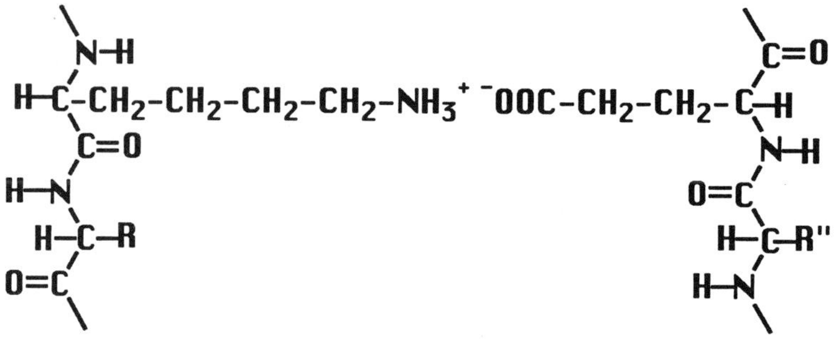

Fig. 7.5 Ionic cross-link in wool

The two functional groups react with each other to form an ionic bond (inner salt formation).

The Macro Structure of Wool

In higher levels of structural organization, strands of alpha helix are twisted or woven together to form large fibrous structures called cables. Several of these cables are packed in bundles to form a microfibril. The microfibrils are then fixed firmly in specific arrangements that lead ultimately to the formation of the macro structure of the wool fiber with all of its different segments.

Of particular interest to the dyer is the shape and nature of the outer layer of the wool fiber. The fiber is rod-like and curled, and becomes gradually thinner from the root end to its tip. The outermost layer, called the epicuticle is in the form of scales pointing in the direction of the tip. Their shape and fineness vary greatly from one type of fiber to another. These scales act as a protective cover for the fiber. Also, their surface is somewhat hydrophobic which makes dyeing wool at room temperature practically impossible. High temperatures and extended time are required to open up the fibers sufficiently so that the dye solution can penetrate into the more hydrophilic inner structure of the wool.

The scaly surface is the main cause for the tendency of wool to felt. In the process of felting, often called 'milling', a wool fabric is rubbed and pulled, usually in the presence of a strong soap solution. During felting, interlocking of the fibers takes place, and the fibers are drawn together and shrink a considerable amount. Therefore, care should be taken when dyeing wool fabrics to prevent undesirable felting and/or shrinking. For example: in a beck dyeing machine, when circulating the fabric through the dye-bath at a fast rate, slipping of the fabric on the reel may cause partial felting and shrinking.

Sometimes unlevel dyeing can result from "tippy dyeing'. The tips of the fibers are exposed to sunlight, thus weathering much more than the sections closer to the root. This exposure causes the tips' scales to break and the tips become more hydrophilic. As a result, irregular dyeing may occur. "Tippy dyeing' is much more pronounced when dyeing with mixtures of dyes (where variations in hue and/or depth of shade between the tips and the rest of the fibers may occur). By proper selection of dyes and dyeing methods, 'tippy dyeing' can be eliminated or at least minimized.

7.2 The Acid Dyes

Most acid dyes are sodium salts of organic sulphonic acids. They consist of an aromatic structure containing a chromogen and a solubilizing group(s), almost always a sulphonic acid salt ($-SO_3^-$ Na^+). The acid dyes are soluble ionic compounds, where the color is contained in the anionic portion. An example of an acid dye is:

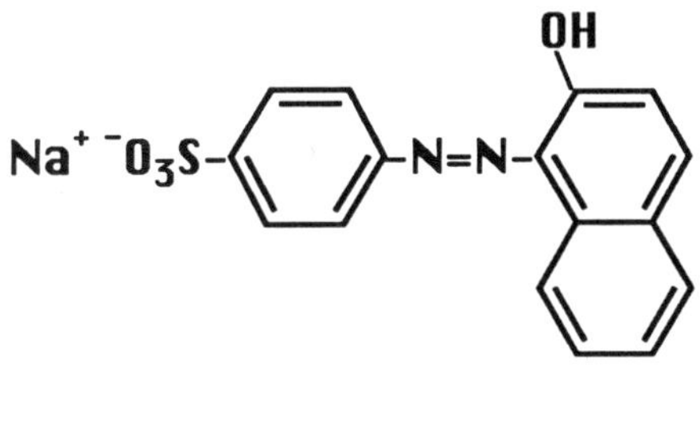

C.I. Acid Orange 7

About two third of the acid dyes contain one or more azo groups as their chromophores. The other acid dyes may contain anthraquinone, triaryl methane, nitrophenyl amine, and other chromophoric derivatives.

Initially acid dyes were so called because an acid was required in their application. The term 'acid dyes' is also consistent with their chemical structure since they are salts of organic acids. Often acid dyes are also referred to as anionic dyes.

Acid dyes are used mainly for both natural and synthetic polyamide fibers (wool, silk, and hair, versus the nylons). All these fibers contain amino groups which serve as the dye sites for the acid dyes. These dyes are also used occasionally in dyeing other fibers such as acid dyeable acrylic and polypropylene.

7.3 The Application of Acid Dyes to Wool

Dyeing Procedure

Exhaust dyeing is the method of choice for dyeing wool and it is carried out at different stages of production: raw stock (loose fibers), slubbing, yarn, fabric, or finished garments. The following is a general dyeing procedure for wool:

The dye-bath is prepared at 49°C (120°F). In addition to the acid dyes, an acid (e.g. 2% H_2SO_4 o.w.f.) or other acidic chemicals are added as required, to obtain the proper pH of the bath. Sodium sulfate (glauber salt: Na_2SO_4, 0-10% o.w.f.) or other dyeing assistants are added as leveling agents. The dyeing machine is run for 5-10 min., after which the wetted material (raw stock, yarn, or fabric) is inserted. The dye-bath temperature is gradually raised (1°C per min.) to the boil while agitating the material and /or circulating the dye solution through the material. Dyeing is continued at the boil for 40-60 min., after which the dyed material is rinsed with cold water and dried. Dyeing time at the maximum temperature can be reduced by half if dyeing is carried out under pressure, at 104-107°C (220°F-225°F). Excessive amount of dyeing time in particular when dyeing above the boil should be avoided to prevent possible damage to the wool fibers.

The purpose of using acids in the dye-bath is to promote exhaustion. Their presence increases both the rate of dyeing and the total amount of dye that can be absorbed by the fiber. The rate of exhaustion is also affected by the concentration of the dye and its substantivity; The lower the substantivity the more acid (a lower pH) is required. Some acid dyes that need strong acids for exhaustion will only stain wool if applied in the absence of an acid. When the amount of acid used is larger than recommended, the dyes may strike fast. With dyes that can not migrate easily, this strike will lead to unlevel dyeing.

Many times acid dyes are synthesized so that their desired pH of application will be as close as possible to the iso-electric point of wool, which is around pH 4.5. At its iso-electric point, wool is in its most stable configuration, where the highest number of ionic cross-links can exist. Moving away from the iso-electric point by increasing or decreasing the pH, will reduce the number of ionic cross-links:

$$\text{-NH}_2 \quad \text{-OOC-} \quad \underset{-H^+}{\overset{+H^+}{\rightleftharpoons}} \quad \text{-NH}_3^+ \quad \text{-OOC-} \quad \underset{-H^+}{\overset{+H^+}{\rightleftharpoons}} \quad \text{-NH}_3^+ \quad \text{HOOC-}$$

Like sodium sulfate other dyeing assistants, such as anionic surfactants, are used occasionally for slowing down the rate of dyeing by competing with the dyes for the dye-sites. Nonionic detergents are also used sometimes to prevent a fast strike. Detergents increase dye solubility, thereby reducing dye substantivity.

Mechanism of Dyeing Wool with Acid Dyes

Initially the acid may react with the wool to produce positively charged ammonium groups ($-NH_3^+$) on the fibers according to either of the following equations:

$$\underset{\text{wool}}{\text{-NH}_2} + HCl \longrightarrow \text{-NH}_3^+ Cl^-$$

$$\underset{\text{wool}}{\text{-NH}_3^+ \; ^-OOC\text{-}} + HCl \longrightarrow \text{-NH}_3^+ \quad HOOC\text{-}$$

The negatively charged colored ions (anions) of the dye are attracted to the positively charged $-NH_3^+$ groups on the fiber, thus combining with the fiber through an ionic exchange reaction to yield the colored fiber:

$$DYE\text{-}SO_3^- \; Na^+ + WOOL\text{-}NH_3^+ \; Cl^- \longrightarrow \underset{\text{(colored fiber)}}{DYE\text{-}SO_3^- \; ^+H_3N\text{-}WOOL} + NaCl$$

In this reaction the substantive dye anion replaces the anion of the acid (Cl⁻, HSO₄⁻, CH₃COO⁻, etc.), which has a very low affinity for the fiber.

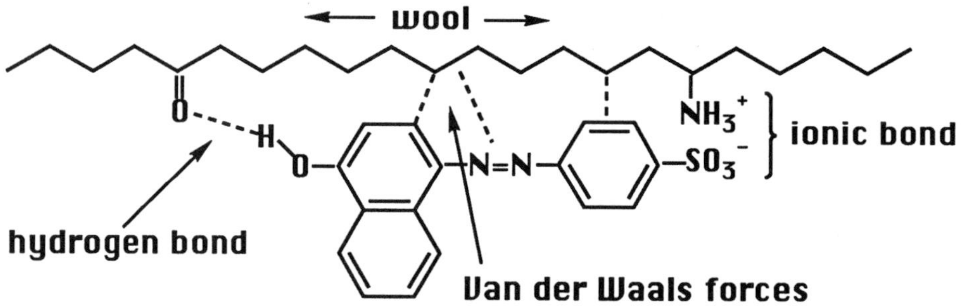

Fig. 7.6 Chemical bonds between an acid dye and wool

The ionic bond formed between the dye and the fiber's macro molecule is the primary bond but not the only one formed between them. Additional bonding takes place between other parts of the colored ion and the fiber (Fig. 7.6). Compared to the strength of the ionic bonds these secondary bonds are very weak and include hydrogen bonds and Van der Walls forces.

The larger the number of these secondary bonds the higher the substantivity. Therefore, larger dye molecules which have more of these second order bonds will have better fastness to wet treatments and laundering.

Sodium sulfate is used as a leveling agent to slow down the rate of dyeing and enhance dye migration. Sulfate ions (SO_4^{--}) compete with the dye anions ($DYE-SO_3^-$) for dye sites. Under normal dyeing conditions, the number of sulfate ions present far exceeds the number of dye ions, and since they are much smaller in size they will diffuse at a faster rate, and initially occupy the dye sites available. The dye ions moving at a much slower rate, will slowly replace the sulfate ions, resulting in a much more uniform dyeing.

7.4 The Different Types of Acid Dyes

The bonds that are formed between wool and the earlier acid dyes are not strong enough to withstand multiple launderings. This limited wet fastness is even more pronounced where milling (felting) operations follow dyeing. In the milling process wool fabrics are felted by mechanically pulling and beating them in presence of a strong soap solution..The need for acid dyes that will not bleed during the milling process resulted in the introduction of the acid milling and super milling dyes. These dyes consist of larger molecules which are capable of forming larger numbers of second order bonds (hydrogen bonds and Van der Waals forces). Though the increased molecular size results in improved wet fastness, these dyes are less soluble in water and are more difficult to migrate.

The subdivision of the acid dyes that follows is based on their methods of application. A comparison of their behavior during application and in normal use is shown in Table 7.1.

Table 7.1 General Trends in Properties of Acid Dyes

Acid Dye	Acid Leveling	Acid Milling	Neutral Dyeing
Type of Acid	strong (2-4% H_2SO_4) C.I. method #3 (pH<3.5)	weak (2-4% CH_3COOH) C.I. method #2 (pH 3.5-5.5)	ammonium salts ($(NH_4)_2SO_4$) C.I. method #1 (pH 5.5-7)
Solubility	high ---> low		
Substantivity	low ---> high		
Level Dyeing	very good --------------------------------> fair		
Rate of Exhaustion	slow --> fast		
Dyeing Time	relatively short ------------------------------> longer time		
Wet-Fastness	fair --->very good		
Color-range	wide range ------------------------------> moderate with bright colors brightness		
Molecular Structure	relatively small size ----------------------> larger size		

7.5 Acid Leveling Dyes

Among the different acid dyes, the acid leveling dyes (equalizing dyes) are the easiest to apply, because they have a relatively small molecular size, and are readily soluble in water. These properties allow these dyes to diffuse and migrate easily through the inner parts of the fibers and level dyeing is readily attained. Examples of these dyes are C.I. Acid Orange 7 (see above), and:

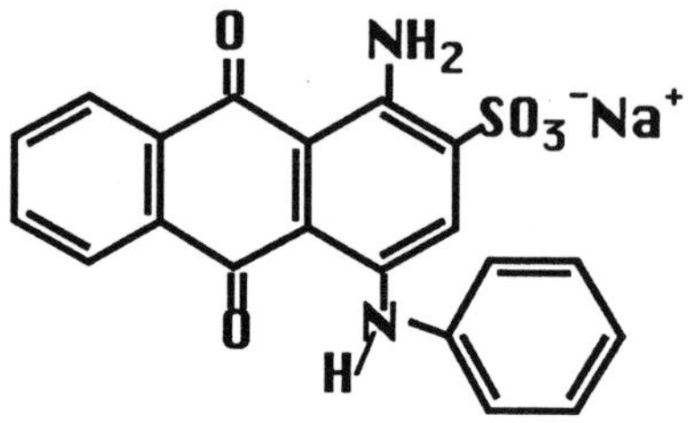

C.I. Acid Blue 25

Dyeing wool with acid leveling dyes requires the use of a strong acid (e.g. 2-4% H_2SO_4 o.w.f.) (C.I. method of application #3, pH of dye-bath below 3.5),and sodium sulfate (about 10% o.w.f.) is used as a leveling agent. The dye-bath temperature can be raised to the boil in a short time. Dyeing is continued for 30-60 min., during which time any unlevelness will be corrected through migration.

The acid leveling dyes are available in a broad range of shades with many bright colors, and their light fastness varies from poor (e.g. some triphenyl methane derivatives) to very good (anthraquinone derivatives). The main disadvantage of the acid leveling dyes, however, is their limited wet-fastness which results from their moderate substantivity and high solubility.

7.6 Acid Milling Dyes

These dyes are made of larger molecules than those of the acid leveling dyes, and require only a weak acid (e.g. acetic acid) for exhaustion (C.I. application method #2, pH of dye-bath: 3.5-5.5). The behavior of the acid milling dyes during dyeing and in normal use is between that of the acid leveling dyes and the neutral dyeing acid dyes. An example of an acid milling dye is:

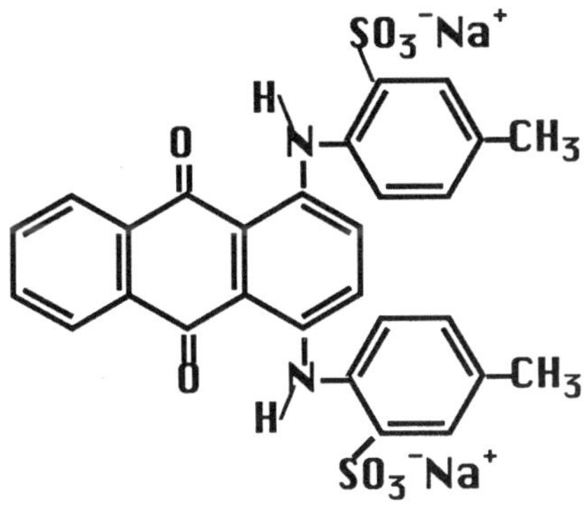

C.I. Acid Green 25

The acid milling dyes have the advantage of being applied at a pH close to the iso-electric point of wool (pH 4.5).

7.7 Neutral Dyeing Acid Dyes

These dyes also called Super Milling dyes, usually have the highest molecular weights, and therefore have very good wash-fastness. An example of these dyes is:

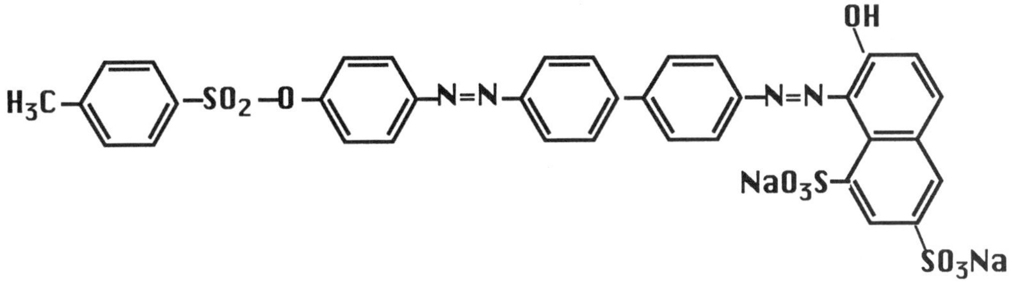

C.I. Acid Red 85

As their name implies, the neutral dyeing acid dys are applied at a pH close to neutral (C.I. Dyeing method #1, pH of dye-bath: 5.5-7). They are not very soluble in water, exhibit a high substantivity, and therefore do not migrate easily. Because of their poor leveling properties the dyes are applied under controlled exhaustion. The dye-bath temperature is raised slowly, in particular during temperature ranges where a small change in temperature brings about a great increase in the rate of dyeing.

The addition of an acid, even a weak acid may force these dyes to strike fast and/or precipitate out of the dye-bath. Therefore dyeing is performed in almost a neutral bath with the use of ammonium salts such as ammonium sulfate or ammonium acetate. At the boil, some ammonia comes off and a small amount of acid is liberated slowly:

$$(NH_4)_2SO_4 \xrightarrow{\Delta} 2NH_3\uparrow + H_2SO_4$$

This will provide but a small amount of acid, enough to promote proper exhaustion.

Usually sodium sulfate is not used with these dyes. Being an electrolyte it may actually increase the rate of dyeing or force these dyes to precipitate. The effect of electrolytes on the solubility of dyes will be discussed in the chapter on Direct Dyes.

Most of the neutral dyeing acid dyes have dull shades. Their larger molecules contain more chromophores, auxochromes, and conjugated double bonds, resulting in increased absorption of visible light along the visible range of the electromagnetic spectrum. To overcome the effect of dulling caused by increasing molecular weight, other dyes have been developed (e.g. Polar dyes by Ciba., and Carbolan dyes by I.C.I.) in which their molecular size is increased by saturated alkyl groups that do not affect the color.

7.8 Chrome Dyes

These dyes, also called mordant dyes, have been used on protein fibers where maximum wet-fastness is required. The chrome dyes have chemical structures similar to acid dyes, but they are capable of forming stable complexes with chromium ions (Cr^{+++}). An example of a chrome dye is:

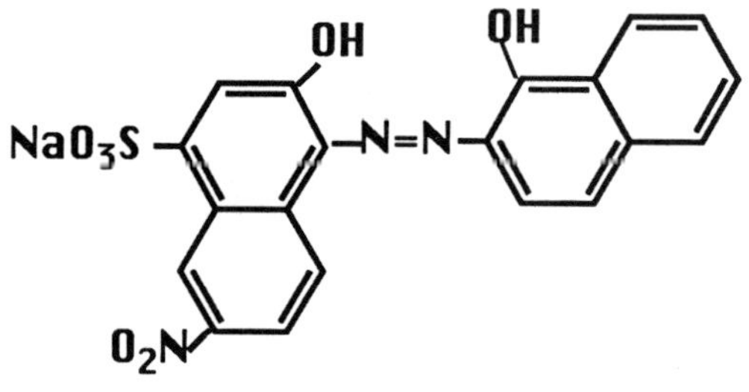

C.I. Mordant Black 11

The typical feature in this structure is the two hydroxyl groups in ortho positions to the azo group. Chrome dyes can be applied separately or together with the chrome compound, and the end result of the dyeing process is the formation of a DYE-CHROME-FIBER complex. A chromium ion can attach to the dye and the fiber simultaneously by ionic and/or coordinated bonds:

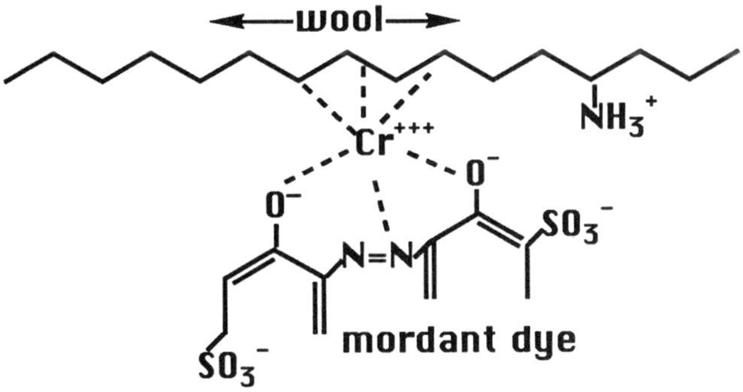

Fig. 7.7 Dye-mordant-fiber complex

A chromium ion may also attach itself to two dye molecules thereby forming a large complex molecule with much better wash-fastness [3].

The chrome compounds used are usually sodium dichromate ($Na_2Cr_2O_7$) or potassium dichromate. In these compounds the chrome is in its 6+ oxidation state. In this state the chrome is not free to form additional bonds with the dye and/or the fiber. During application, however, the Cr^{6+} is reduced to the more stable Cr^{3+}, which then acts as the complexing agent. The wool has reducing sites (e.g. cysteine groups that can be oxidized to form cystine linkages) that allow the reduction of the chrome to take place. The reduction is enhanced by carrying out the application in presence of sulfuric acid. In acidic solutions, chrome is most stable in its 3+ oxidation state.

Chrome dyes can be applied by the following methods:
1. Bottom Chrome: The chrome compound is applied first after which, in a separate bath the dyeing with the chrome takes place.
2. Meta Chrome. Both the chrome compound and the chrome dyes are applied from the same bath.
3. Top Chrome, or After Chrome. The dyes (chrome dyes) are applied first by one of the methods used for regular acid dyes. Then the chrome compound is applied as an after-treatment. The chrome compound can be applied from the same dye-bath or from a new bath.

Of the three methods the first two are now of historical value. The Top Chrome method is being used, however, to a large extent, when dull colors with excellent wash-fastness are desired. The following is a typical application of chrome dyes by the Top Chrome method:

The dye-bath is prepared with the proper amounts of an acid (sulfuric or acetic acid) and sodium sulfate, and the chrome dyes. The fibers are introduced and the dye-bath temperature is raised to the boil in 30-45 min. Dyeing continues at the boil for additional 30-45 min. During this time a small amount of acid is added if needed to complete exhaustion. Then the dye-bath temperature is dropped to 71°C (160°F) by removing part of the dye solution and replacing it by the same amount of cold water. Then the proper amount of a solution of the chrome compound is added (e.g. for 1% dyeing, 0.5% sodium dichromate is used). The temperature of the bath is raised again and kept at the boil for 30-60 min. The treatment is completed by rinsing and drying.

In addition to their excellent wash and crock fastness, chrome dyes have very good good light-fastness, and very good migration properties.

The main disadvantage of the chrome dyes is that during complexing with chrome a change in hue takes place. Combined with the Cr^{3+} ion the visible light absorption of the dye is significantly changed. Since the color change takes place when the dye becomes firmly attached to the fiber, color matching in this method is difficult. Other disadvantages associated with the Top Chrome method are the dulling effect of the chrome on the original color, the lengthy time of the dyeing procedure, and the toxicity of the chrome compounds (mainly in their higher oxidation state).

7.9 Premetallized Acid Dyes (Metal Complex Dyes)

The next step in the development of acid dyes with superior wash-fastness has been the incorporation of a transition metal ion in the dye molecule by the dye manufacturers. These dyes usually contain a Cr^{3+} ion (some contain a cobalt ion) bonded to one (1:1 complex) or two (2:1 complex) dye molecules by ionic and and coordinate bonds. Examples of these dyes are C.I. Acid Blue 158 (1:1 complex), and C.I. Acid Black 60 (2:1 complex):

Note that in both complexes, though the Cr^{3+} ion carries three positive charges, the net charge of the colored ion is negative. Therefore, they can behave as acid dyes.

In addition to very good wash-fastness, the light-fastness of the dye is also improved by the presence of the metal. However, the two types of premetallized dyes, the 1:1 metal complex and the 2:1 metal complex, differ greatly in their method of application.

C.I. Acid Blue 158   (1:1 metal complex)

C.I. Acid Black 60   (2:1 metal complex)

The 1:1 metallized dyes bond to the fibers not only by ionic and second order bonds, as do the regular acid dyes, but also by coordinate bonds through the Cr^{3+} ion. These dyes are available in a wide range of colors of moderate brightness. The 1:1 metallized dyes are easy to dissolve, and have good leveling properties. Their main disadvantage, however, is that they require a large amount of acid (approximately 8% H_2SO_4, pH 2) for exhaustion. This strong acidity may cause damage to the wool. Therefore, the material has to be checked in advance to see whether it can be dyed by this method at all.

The 2:1 metallized dyes, because of their large molecular size, behave similarly to the neutral dyeing acid dyes. Their application procedures are practically the same as those used for the neutral dyeing acid dyes (C.I. method #1, pH 5.5-7).

The 2:1 metallized dyes are noticed for their very good wash-fastness and light-fastness. Another advantage of these dyes is that when using several different colors in the same bath they tend to exhaust at the same rate, and to the same extent. This is very important when using them on nylon, where they are used extensively. The main disadvantages of the 2:1 metal dyes are the lack of bright shades, high cost, and poor migration.

Another group of acid dyes closely related to the 2:1 metallized dyes are the sulfonated 2:1 metallized dyes. These dyes, offered by several dye manufacturers (e.g. the lanaset dyes of Ciba-Geigy and the Lanasyn S dyes of Sandoz) are made by adding sulphonic acid groups to 2:1 metallized dyes. The main advantage of these dyes is that they are applied at a pH of 4.5-5, near the iso-electric point of wool; otherwise they behave similarly to the 2:1 metallized dyes, both during application and in normal use.

7.10 Stripping Acid Dyes From Wool

Partial stripping can be accomplished by treating the dyed material at the boil with a bath containing 20 to 30% Glauber salt, for about one hour. This treatment is safe and used to remove just small amounts of dyes. For removal of larger amounts a very slightly alkaline bath can be considered. A small amount of ammonia can be used but attention must be given not to exceed pH 8-8.5, in order to avoid any harm to the wool due to excess alkali.

Complete removal of certain acid dyes can be done with Zinc-sulphoxylate-formaldehyde. This reducing agent requires an acidic pH which is favorable to wool. The reduction of the dyes is carried out at or near the boil at a pH of 4.5-5, for 20-30 min.

References

1. The Theory and Practice of Wool Dyeing, Fourth Edition, 1972, C.L. Bird, The Society of Dyers and Colourists, Bradford, Yorkshire, England.
2. Wool, A SANDOZ Manual, Sandoz LTD, Basle, Switzerland, 9012/80.
3. Rowe F.M. et al, J.S.D.C., 1946, **62**, 372

Chapter 8.

DYEING CELLULOSE WITH DIRECT DYES

8.1 Cellulose

Cellulose is the principle building material of the vegetable world. Cotton, linen, and jute, are examples of natural cellulosic fibers, differing only in their geometric molecular structure. Rayon (or viscose) is a regenerated cellulosic fiber manufactured from wood.

Cellulose is a linear polymer made of anhydro-glucose units:

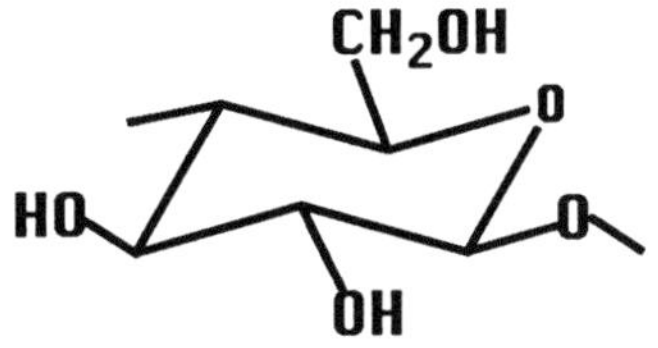

Anhydro-glucose Unit

bonded to each other through acetal linkages:

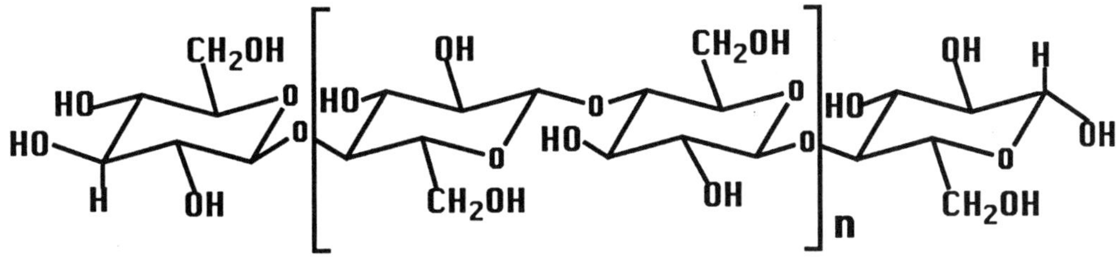

CELLULOSE

There are usually between 2000 and 4000 anhydro-glucose units in the polymeric molecule of cotton, however, in certain grades of cotton the number of anhydro-glucose units per molecule may be as high as 10000. The degree of polymerization decreases significantly when cotton fibers go through preparation treatments with basic solutions. In rayon, the number of anhydro-glucose units per macromolecule is less than 500, whereas in high tenacity rayon, and polysonic fibers, the number is slightly higher than 500.

In cotton, it has been found that about two thirds of its structure is in the crystalline form, and one third of it is in the amorphous form. When placed in water there is a volume increase of 40%, as well as an increase in the tensile strength of the fibers. In rayon only one third of its structure is crystalline and the rest is in the amorphous form. When immersed in water, rayon swells to about the same degree as cotton but its tensile strength decreases.

The characteristics of cellulosic fibers are primarily due to their huge number of hydroxyl groups. There are three hydroxyl groups present in each anhydro-glucose unit of cellulose. The cellulosic fibers are, therefore, hydrophilic, with a high moisture regain. The comfort properties associated with cotton fiber are mainly due to their hydrophilic nature, but also because of their fine structure. Cotton can have up to 15% moisture content without feeling wet.

In the crystalline regions of cellulosic fibers, the polymeric chain molecules are lengthwise bonded to each other through hydrogen bonds, formed between hydroxyl groups of neighboring chains. Water molecules cannot enter the tightly packed net work thus obtained, and therefore the fibers remain insoluble in water. The swelling of cellulosic fibers when immersed in water is the result of water molecules entering the amorphous regions, forcing the polymeric chains to move further apart. The water molecules inside the cellulosic fibers are bonded to them chemically, mainly through hydrogen bonds. The hydroxyl groups of cellulose also serve as bonding sites for certain dyes and chemical finishes as will be shown later.

Cellulose fibers are sensitive to acids. Treatments of cotton with diluted acid solutions at high temperatures may cause damage (see also: 'Carbonizing of wool'). On the other hand cellulosic materials are quite stable to alkali solutions. Some degradation, shown by a small loss in weight, takes place in prolonged alkali treatments at high temperatures, as in 'Caustic Scouring'.

Strong oxidizing agents may damage cellulosic materials. However, cold diluted solutions of oxidizing agents are used in bleaching and dyeing of cellulosic fibers. Hydrogen peroxide is the only oxidizing agent used for cellulose at high temperatures. Reducing agents such as hydrosulphite ('hydro') do not affect cellulosic fibers and are used on them in dyeing and in stripping of dyes.

8.2 The Direct Dyes

Similar to acid dyes, direct dyes are also soluble anionic dyes. They too consist of an aromatic structure containing a chromogen and solubilizing groups. The main difference between them, however, is that direct dyes are substantive to cellulosic materials, whereas acid dyes may only stain them. The direct dyes are so called because they were the first dyes to dye cellulosic fibers directly, without the need for a pre-treatment of the fibers with a mordant. Their application is rather simple since the only dyeing assistant usually used is a strong electrolyte such as sodium chloride or sodium sulfate, to assist in exhaustion.

The majority of the direct dyes are disazo and trisazo derivatives. Examples of azo direct dyes are:

C.I. Direct Red 81   (Class A)

C.I. Direct Black 22   (Class C)

Many direct dyes are soluble copper complexes. They are noted for their very good light fastness. An example of a copper complex direct dye is C.I. Direct Blue 218 (see below).

Many of the earliest direct dyes, including the first direct dye Congo Red (synthesized by Boettiger in 1884) were derivatives of benzidine, which was found to be carcinogenic, and the production of these dyes in the U.S.A. has been discontinued. However, once the benzidine is incorporated into a direct dye it is no longer toxic, and therefore some of these benzidine derivative dyes are imported.

Copper Complex of:

$$\text{C.I. Direct Blue 218} \qquad \text{(Class A)}$$

Note that since direct dye molecules have all the components of acid dyes, they could be used as such. Indeed, some direct dyes are recommended to be used for polyamides and protein fibers. Because of their large molecular size and lower solubility as compared to regular acid dyes, they are applied to these fibers under similar conditions to those used in the applications of neutral dyeing acid dyes. On the other hand, a neutral dyeing acid dye may not dye cellulose at all, unless it happens to have the required substantivity to cellulose. Prior to the appearance of the synthetic fibers, cotton/wool blends were popular. Certain direct dyes were used to dye these blends into a solid shade (union dyeing). It seems that lately, because of fashion and comfort, cotton/wool blends are gaining new interest.

The mechanism by which direct dyes become attached to cellulose is assumed to be through the formation of a large number of weak attractions between the dye and the fiber. These weak attractions consist of Van der Waals forces, and hydrogen bonds. Ionic bonds such as those that exist between acid dyes and wool are not possible here since cellulose is not ionic in nature. Accordingly the bonding of direct dyes to cellulose is not sufficient to yield good wash-fastness.

Since the first direct dyes had distinctive structural features of being long, linear, and planar, it was believed that these three properties were essential for a dye to be substantive to cellulose. Indeed, such a structure permits the molecule to enter the fiber easily, and align itself along a polymeric chain, so that it will have a large number of attachments with the fiber. An example of such a structure is the above C.I. Direct Black 22. Since then, however, many other direct dyes have been developed which do not possess all the three features; yet they are all recognized by their large complex structure. Examples of such dyes are: C.I. Direct Green 26, C.I. Direct Green 28, and C.I. Direct Blue 86.

C.I. Direct Green 26   (Class B)

C.I. Direct Green 28   (Class B)

8.3 General Properties of Direct Dyes

Direct dyes provide a simple and relatively inexpensive way of dyeing cellulosic fibers. The main disadvantage of direct dyes is their limited wash-fastness. Since direct dyes are attached to cellulose through weak bonds and are water soluble they tend to come out during wet treatments. Therefore, they are used frequently on materials that are not washed at all, such as ribbons for gift wrapping or paper. To improve their wash-fastness, different after-treatments have been developed. Nevertheless, the improvements obtained are many times still insufficient to satisfy the usual fastness requirements. Therefore, direct dyes are used today mainly for producing light shades.

Direct dyes are available in all shades and in large numbers, but there are not too many bright colors. The majority of the direct dyes are the azo dyes (bis and tris azo) with extended conjugated double bonds system and a large number of auxochromes. These types of structures have an excessive number of absorptions in the visible region, and therefore appear as dull colors. Among the bright colors available are bright greens, however, they are more expensive than the others.

An example of a direct dye with a bright green color is C.I. Direct Green 26. In its structure there are two chromogens that are not connected to each other through conjugated double bonds, and therefore act independently. The chromogen with the azo group between the two phenyl groups absorbs blue light, and the other chromogen with the two azo groups absorbs red light. The result is that when exposed to visible light, this dye reflects only green light. Another example of a dye with a bright green color is C.I. Direct Green 28. In this dye there are also two separated chromogens. The chromogen with the anthraquinone group absorbs red light, and the chromogen with the azo group absorbs blue light.

Light fastness of direct dyes varies widely from poor to very good. Some direct dyes are metallized with copper to increase their light-fastness; in other cases, copper salts are applied as an after-treatment for improving light-fastness and wash-fastness.

As with light-fastness, fastness to other conditions such as exposure to high temperatures, chlorine, and perspiration, varies greatly from one dye to the other, but there are many direct dyes to choose from in order to find the proper ones for a specific end-use.

8.4 The Application of Direct Dyes

Direct dyes are applied by both exhaust and continuous methods. In exhaust dyeing the dyes are usually applied at temperatures near the boil. There are, however, direct dyes that are applied at temperatures as low as 66ºC (150ºF). In continuous dyeing they are usually applied by the pad/steam method.

In a typical exhaust dyeing the direct dyes are dissolved carefully, filtered or strained, and then added to the dye-bath. The cellulosic substrate is placed in the dye-bath at 49ºC (120ºF), and dyeing is continued at this temperature for 15-20 min. Then common salt or glauber salt in the solid form is added portion-wise, over a period of 45-60 min., while the temperature is slowly raised to the maximum temperature of dyeing. Additions are made by starting with a small amount of salt followed by larger ones. For example, the salt to be added is divided into four portions: 1/8, 1/8, 1/4, and 1/2 of the total amount.

The total amount of salt used is usually in the range of 5-25%, depending on the depth of shade, liquor ratio, and the type of direct dyes used. If the shade is satisfactory, the dyed material is then rinsed with cold water. At this stage an after-treatment may follow to improve the wash-fastness of the dyes.

Effect of Electrolytes

The addition of electrolytes to the dye-bath promotes the exhaustion of direct dyes. This is explained (T. Vickerstoff, The Physical Chemistry of Dyeing, Interscience Publishers, 1984, p. 231) as follows: "The cellulosic fibers, when immersed in water, assume a very slightly negative charge. This negative charge makes it difficult for the colored anion of the direct dye, which is also negatively charged, to penetrate into the fibers. When an electrolyte (e.g. NaCl) is added, some of its positive ions (Na+) are attracted to the fibers and neutralize their negative charge. The negatively charged colored ions will now diffuse into the neutral fibers."

The neutralizing effect is not the only reason for the ability of the added salt to drive the direct dye into the fibers. Equally important, however, is to realize that the introduction of a strong electrolyte to a dye-bath significantly reduces the solubility of dyes in the bath. The polar water molecules prefer to mix with the charged particles of a strong electrolyte rather than with dye ions which are partially hydrophobic. As a result a substantial amount of water is occupied by the electrolyte as water of hydration, thus reducing the amount of water available for keeping the dye in solution. Obviously, for this reason when dyeing at high liquor ratios, larger amounts of salt are required for exhaustion. A larger amount of salt is also used when dyeing dark shades.

Though sodium chloride is less expensive, sodium sulfate is more often used as the electrolyte since it does not cause as much corrosion of equipment as sodium chloride does.

Effect of Solubility

Though the direct dyes are water soluble, in solution they tend to form aggregates which is typical of dyes that are not very soluble. The aggregated particles may consist of a few molecules to clusters of up to twenty and more dye molecules. These aggregates are too large to diffuse in to the fibers. However, the aggregates are in a state of a dynamic equilibrium with the single dye molecules in the bath. As soon as the available single dye molecules are absorbed by the fibers, the equilibrium is disturbed, and aggregates will dissociate to furnish more single molecules.

As the concentration of the dye increases so does the degree of aggregation, and this may eventually lead to the precipitation of the dye. Therefore, dyeing dark shades at low liquor ratios should be avoided. Dye aggregation also increases with addition of electrolytes. However, aggregation of dye molecules decreases with increasing dye-bath temperature.

When dyeing at low liquor ratios as in continuous dyeing, direct dyes with good solubility should be used.

Effect of Temperature

The absorption of direct dyes by cellulosic fibers increases with increasing temperature until a maximum is reached. Above this temperature the absorption of many direct dyes decreases. The temperature of maximum exhaustion for a specific direct dye may be as low as 49°C (120°F) or as high as the boiling temperature. Typical behavior curves of three direct dyes: x, y, and z, are displayed in Fig. 8.1.

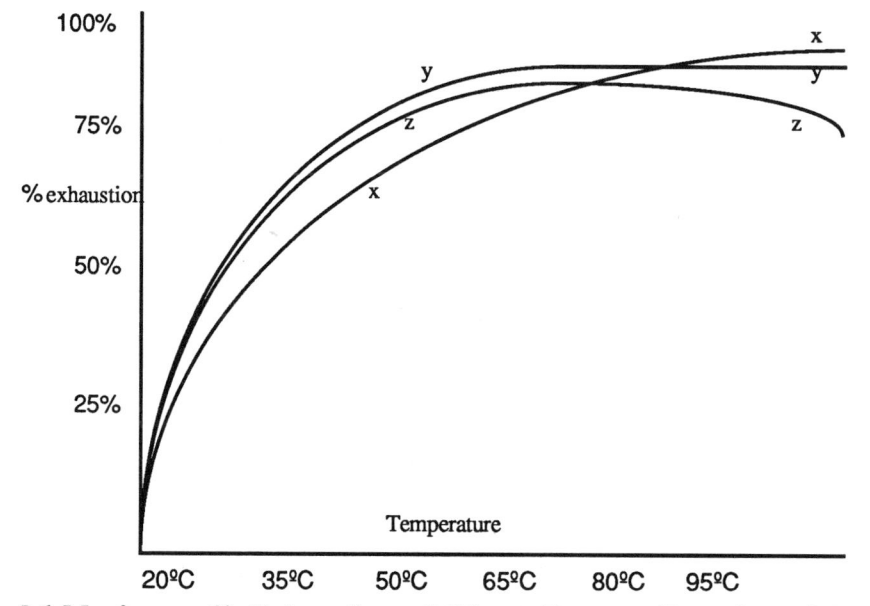

Fig. 8.1. Maximum % Exhaustion of Direct Dyes as Function of Temp.

In contrast, the rate of dyeing continuously increases with increasing temperature. Other dyeing properties that are also continuously enhanced with increasing temperature include both uniformity of dyeing and the extent of dye penetration.

For best results, whenever possible, dyeing procedures are arranged so that the temperature is maintained above the maximum exhaustion temperature during most of the dying time. For example, when dyeing with dyes that have a maximum exhaustion temperature around 82°C (180°F), the dye-bath temperature is kept near the boil most of the time. This high temperature will provide better leveling and penetration results, and a shorter dyeing cycle. Toward the end of the dyeing application the temperature is lowered to 82°C (180°F) and kept there for a short time to allow for maximum absorption of the dyes.

Note that when the dyes in a combination have significantly different maximum exhaustion temperatures, care must be taken not to miss the correct shade due to 'late exhaustion'. This could happen when dyeing is carried out near the boil and one of the dyes has a maximum exhaustion temperature at a significantly lower temperature (e.g. dye z in Fig. 8.1). In this case, after dyeing near the boil, the sample taken out of the dye-lot has the correct shade. However, upon cooling, the dye with a maximum exhaustion at a lower temperature will continue to exhaust. Therefore, to prevent 'late exhaustion', after dyeing, the dye-bath should be dropped while hot, before the cooling step. Obviously the best way to choose, when possible, is to select dyes with similar exhaustion curves.

Effect of pH

Direct dyes are usually applied from a neutral dye-bath. Sometimes, the dyeing is carried out under slightly alkaline conditions using a small amount of sodium carbonate (1-2% o.w.f.), or a similar amount of trisodium phosphate. Applying direct dyes from a slightly basic solution will yield better leveling results (improved dye solubility and/or reduced rate of exhaustion), and will protect direct dyes that are sensitive to acidic conditions.

There are however, many direct dyes that will decompose when applied from a basic dye-bath at high temperatures and for prolonged periods of time.

In particular, the pH sensitivity of the direct dyes in question should be checked carefully when dyeing above the boil. An example is the case of dyeing polyester/cotton blends under pressure from a single dye-bath containing the dyes for both fibers. Here the dyes are applied from an acidic dye-bath (a pH of around 5 for the protection of the disperse dyes), and the polyester is dyed first with the disperse dyes at temperatures in the range of 115°C-130°C (239°F-266°F). Then the temperature is lowered to the boil or below (e.g. 60°C (176°F) for direct dyes with maximum absorption near this temperature) and kept there for a short time to complete the dyeing of the cotton with the direct dyes.

8.5 Classification of Direct Dyes

Direct dyes vary greatly in their behavior during the dyeing process. Therefore, they are classified according to their method of application. The commonly practiced method, suggested by the Society of Dyers and Colorists in 1946, and currently used in the Color Index, divides the direct dyes into three groups:

Class A: Dyes with very good leveling or migration properties. These dyes require large amounts of salt for exhaustion.

Class B: Dyes with fair leveling properties which have to be applied by means of controlled additions of salt.

Class C: Dyes with poor leveling properties, which have to be applied under carefully controlled conditions. Both, the additions of salt and the raising of temperature has to be controlled. These dyes are applied with the least amount of salt or no salt at all.

When selecting compatible dyes for a specific application attention should also be given to their maximum exhaustion temperature.

8.6 After-Treatments of Direct Dyes

Direct dyes are often after-treated mainly to improve their wash-fastness; sometimes also to improve their light-fastness.

The most common after-treatment agents used to day are the cationic fixing agents, such as an alkyl-trimethyl-ammonium phosphate. The cations of these agents combine with the colored anions of the direct dyes to form large complex molecules with low solubility in water. The enlargement of the dye molecule inside the fiber makes it also more difficult for the dye to come out of the fibers. However, the cationic fixing agent may also have an adverse effect by causing the direct dye to move toward the surface of the fibers. Indeed the effect of cationic fixing agents on the wash-fastness of different direct dyes varies greatly from a significant improvement to no improvement at all. When choosing a proper after-treatment it should also be taken into consideration that many times cationic agents have a dulling effect on the original shade; in some cases a change in hue takes place. Sometimes a significant reduction in light-fastness may also occur.

In a typical after-treatment the bath is prepared with the proper amount of the cationic fixing agent (e.g. 1-2% o.w.f. of the fixing agent for material dyed with 1% of a direct dye), and a small amount of acetic acid (pH of bath: 5-6). The dyed material is placed in the bath, and treated at 49-54°C (120-130°F) for 15-20 min. Then the treated material is rinsed with cold water and dried.

The light-fastness of some direct dyes can be improved by an after-treatment with copper sulfate. These dyes are capable of forming stable complexes with copper ions. A small improvement in wash-fastness is also observed. Some products offered for after-treatments are mixtures of copper sulfate and a cationic fixing agent for the improvement of both light and wash-fastness.

Because of harmful ecological effects, many transition metal compounds, including copper compounds, are subject to effluent discharge regulations. In order to minimize the amount of copper in the effluent, the copper sulfate in the application bath should not be present in a large excess. It has been shown, however, that the required amounts of copper sulfate for specific applications are far below the recommended traditional amounts [2].

Formaldehyde has been found quite effective in improving wash-fastness of certain direct dyes. Though not used by itself because of toxicity reasons, formaldehyde derivatives may be found in commercial products, mixed with cationic fixing agents to improve their performance. The effect of formaldehyde is observed many times when the wash-fastness of direct dyes is significantly improved by permanent press finishes that follow dyeing. These treatments involve the use of formaldehyde derivatives which may also bond to some extent with direct dyes.

8.7 Stripping of Direct Dyes

Stripping of direct dyes that were not after-treated is relatively easy. Partial stripping can be done by treating the dyed material at the boil in a bath containing a small amount of sodium carbonate (e.g. 1g. per liter). Complete removal of the dyes is usually done by reducing them at the boil with hydrogen dithionite ('hydro'), or by bleaching with sodium chlorite.

Stripping of direct dyes that were after-treated requires first the removal of the after-treatment agent. Cationic fixing agents are removed by a treatment with an anionic detergent (1-2g./liter) at the boil for 20-30min. Metal ions, such as copper ions from after-treatments, or from metallized direct dyes, are removed by treatments with sequestering agents. A complete stripping in this case will consist of two steps: the dyed material is treated at the boil with a sequestering agent such as EDTA (e.g. 1-2g./liter), after which the material is treated in a second bath with a reducing or an oxidizing agent.

8.8 Other Classes of Dyes for Cellulosic Fibers

In addition to direct dyes, other classes of dyes have been developed for coloring cellulose. These includes: azoic dyes, vat dyes, sulfur dyes, and reactive dyes, and each group provides some attractive features. The main advantages as well as disadvantages of these dyes will be discussed in the following chapters.

Note that during the dyeing stage of theses dyes, a substantive anion (colored or not) is applied similarly to the application of direct dyes.

References

1. T. Vickerstoff, The Physical Chemistry of Dyeing, Interscience Publishers, 1954.
2. Use of Heavy Metals in Textile Wet Processing., A.A.T.C.C., Book of Papers, October 1989.
3. The Dyeing of Cellulosic Fibers, S.R. Cockett, and K.A. Hilton, 1961, Ebenezer Baylis and Son, Limited the Trinity Press, Worcester and London, England.
4. J.R. Aspland, The Application of Anions to Nonionic Fibers, Cellulosic Fibers and Their Sorption of Anions, Textile Chemist and Colorists, Vol. 23, October 1991, p. 14

Chapter 9

AZOIC DYES

The azoic dyes, also called naphthols, are insoluble azo dyes synthesized by the dyer on the fiber. This method consists of applying two relatively small soluble components, one at a time, that combine inside the fiber to produce a larger insoluble colored molecule: the azoic dye.

The first component, called the naphthol or the coupling component, is mostly a beta naphthol derivative. The naphthols, carrying the weakly acidic hydroxyl group (phenol type), are made soluble by reacting them with alkali to form their salt (the naphtholate):

$$\text{a naphthole} + \text{NaOH} \longrightarrow \text{the naphtholate}$$

The naphtholate is then applied in a manner similar to that of the application of direct dyes. The R-group attached to the naphtholate must be large enough so that the molecule will have the required substantivity to cellulosic fibers. Next, the second component, the diazonium salt, is prepared by reacting an aromatic primary amine, referred to as the base or the diazo component, with nitrous acid (HNO_2). This acid is very unstable and prepared in situ by mixing $NaNO_2$ with HCl or H_2SO_4:

$$R\text{-}C_6H_4\text{-}NH_2 + NaNO_2 + 2HCl \longrightarrow R\text{-}C_6H_4\text{-}N\equiv N^+\ Cl^-$$

diazotization → the diazonium salt

The fibers are then treated with the diazonium salt solution and the coupling reaction takes place; the two components combine to yield the dye:

the coupling reaction

Note that the dye molecule thus formed is non-ionic, and does not contain a solubilizing group.

A summary of the chemical reactions involved in the application of azoic dyes is shown with a specific combination in Fig. 9.1. To avoid confusion note that the components used in the process are referred to by several different names, all describing the nature of the component or its role in the dyeing process.

Since there are more than 40 coupling components and about the same number of diazo components available, theoretically one can produce close to 1600 combinations. Out of these a small number of combinations, still enough to choose from, are recommended by dye manufacturers.

9.1 General Properties of Azoic Dyes

The azoic dyes are used primarily for coloring cellulosic materials. They can also be applied, with proper modifications, to hydrophobic fibers such as polyester and triacetate. Azoic dyes are used sometimes on hydrophobic fibers to produce black shades.

Azoic dyes have overall very good fastness, in particular fastness to multiple launderings and sunlight. Therefore they are suitable for out-door products (flags, draperies, tents, etc.). Light-fastness of light shades, however, may not be adequate. The main benefit in using azoic dyes is that very good wash-fastness on cellulosics is achieved at a reasonable price.

Though the azoic dyes are noted for their full range of red colors along with maroons, scarlets and burgundies, they have color limitations. The azoic dyes lack green and bright blue colors.

Another problem often associated with azoic dyes, in particular when dyeing heavy shades, is their tendency to have poor crocking (rubbing)-fastness. This is the result of dye-formation, during the coupling reaction, on the surface rather than inside the fibers. To avoid crocking, dyeing procedures must be followed carefully, and the dyed fibers

The Naphthol

2-hydroxy-3-naphanilide
Type: beta naphthol derivative
C.I. Azoic Coupling Component 2
Commercial name: Naphthol AS

(insoluble)

The Base

p-nitro-aniline
Type: aromatic primary amine
C.I. Azoic Diazo Component 37
Comm. name: Fast Red GG Base

(insoluble)

↓ NaOH

↓ (diazotization) 1) $NaNO_2$ 2) HCl

The Naphtolate

Type: Sodium salt of beta-naphtol derivative

(soluble)

The Developer

p-nitro-bezene-diazonium chloride
Type: a diazonium salt
Commercial name: Fast Red GG Salt

(soluble)

coupling

The Azoic dye

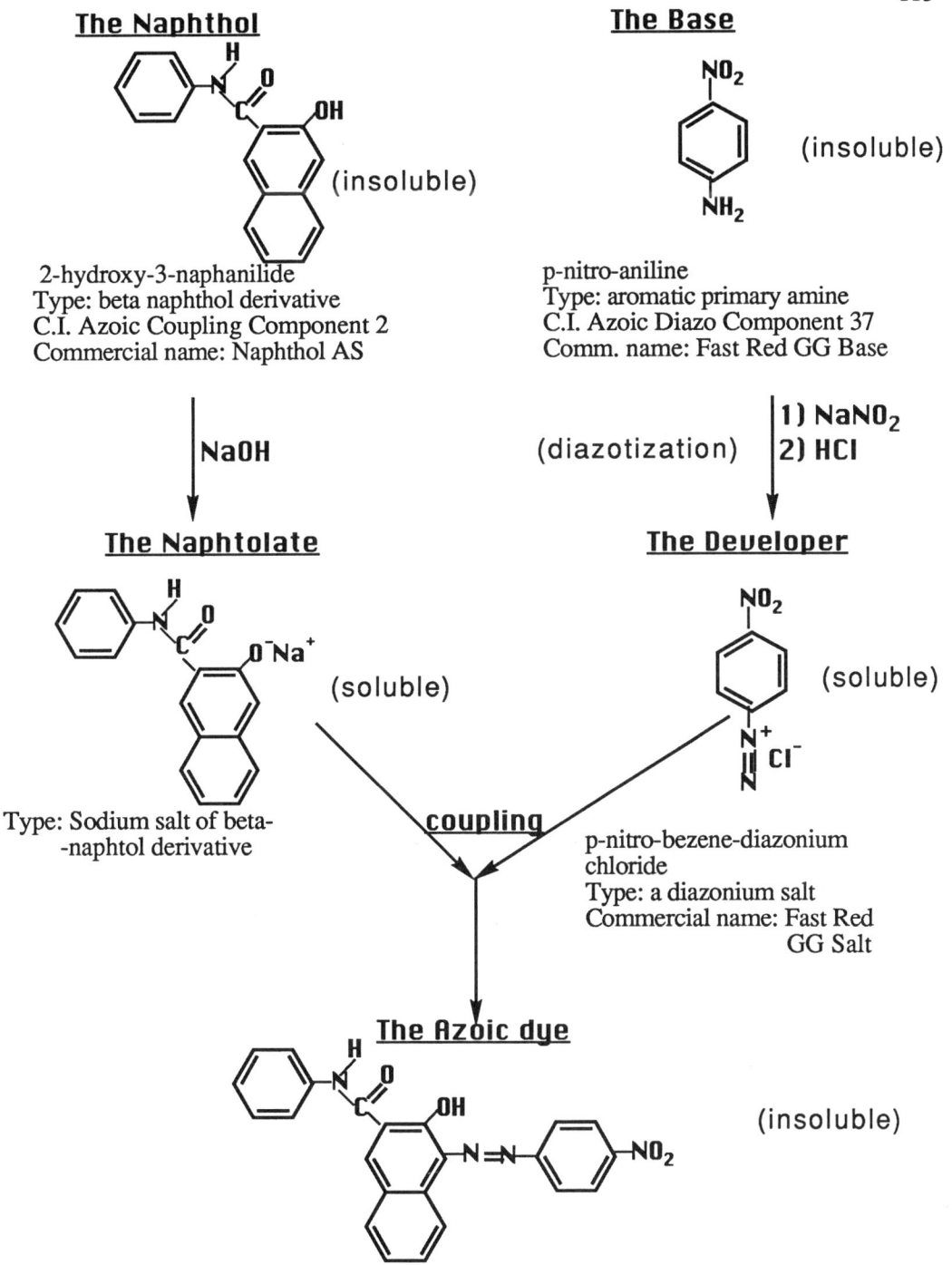

(insoluble)

Fig. 9.1 Formation of an Azoic Dye by the Coupling Reaction

should be washed (after-washed) properly. In general, better wash-fastness and crocking-fastness are obtained with rayon as compared to cotton.

The use of azoic dyes has been declined lately, mainly because of the complexity and time consuming nature of the application, but also because of color limitations. To simplify the dyeing procedures, dye-manufacturers also supply the naphthol and the base in their soluble form, i.e. as the naphtholate and the diazonium salt. As a group of dyes for cellulosics, the azoic dyes still remain the preferred choice for attaining a full depth of bright red ,orange, and yellow colors with a high standard of fastness.

9.2 The Application of Azoic Dyes to Cellulosic Fibers

The application of azoic dyes, by exhaust or continuous processes, consists of the following major steps:
1. Naphtholation - Application of the naphtholate(s)
2. intermediate step - Removal of excess naphtholate from surface of fibers
3. Coupling - Application of the diazonium salt to fibers
4. After-treatment - Thorough cleaning and soaping of the dyed substrate

Exhaust Dyeing.
1. Naphtholation.
First, a solution of the naphtholate is prepared. The following are two common methods used:

Cold dissolving method. Paste the naphthol with a small amount of alcohol (both ethanol or methanol can be used). Add the required amount of NaOH in the form of a concentrated solution until a clear solution is obtained. The amount of NaOH used varies greatly from one naphthol to the other and is in the range of 0.2 to 1.0 Kg of solid NaOH per 1.0 Kg of the naphthol. The concentrated solution is then diluted to the desired volume.

Hot dissolving method. Paste the naphthol with Turkey Red oil or sulfonated oil, and a small amount of water. Raise the temperature to ~85°C (185°F). Add the required amount of alkali in the form of a concentrated solution. Continue to heat for a few min., while stirring, and raise temperature to the boil if needed until a clear solution is obtained. The concentrated solution is then diluted to the desired volume.

Naphtholates in solutions are not very stable, and therefore it is best to prepare them immediately before use. A light precipitate observed during the application can be overlooked. Once a heavy precipitate is formed the solution should be discarded, and a fresh bath is prepared.

In exhaust dyeing, the naphtholate is applied at temperatures between 20°C (68°F) and 50°C (122°F), for 30-40 min., according to dye-manufacturer's recommendations. Naphtholates tend to form insoluble salts with polyvalent metals, therefore, EDTA or a similar sequestring agent is usually added to the dye-bath.

2. Intermediate step

The purpose of this step is to remove, from the fiber's surface, as much of the naphtholate solution as possible in order to avoid crocking problems. This is done by centrifuging, squeezing, vacuum extraction, or other means of hydroextraction. Typically, when dyeing with naphthols of low substantivity, the fabric is dryed. When dyeing with naphthols of high substantivity, however, a salt rinse may be sufficient. This treatment consists of rinsing the naphtholated fibers with a salt solution (as much as 30g. salt per liter) containing a small amount of Na_2CO_3. The base is added to keep the naphtholate from changing back to its original insoluble form. This salt rinse will drive the naphtholate further into the fibers. It will also leave some of the salt on the surface of fibers which will prevent the naphtholate from coming out of the fibers when immersed in the diazonium salt solution.

3. Coupling

The diazonium salt is prepared as mentioned above by reacting the primary aromatic amine (the base) with a mixture of $NaNO_2$ and HCl (or H_2SO_4). Three moles of acid are used for each mole of the base. Two moles are consumed during the formation of the diazonium salt, and one mole of the acid serves to keep the pH of the solution acidic, which is necessary for the coupling reaction to occur.

The bases themselves are stable and can be stored for long periods of time without changing. Regular diazonium salts, however, are unstable. They can be, however, prepared in a stabilized form and purchased as such. An example of a stabilized diazonium salt is a complex made by combining the diazonium salt with zinc chloride: $[R-NN]_2ZnCl_4$. Obviously it is much easier to start with the stabilized diazonium salts. They are easy to dissolve and ready for use. They are, however, much more expensive than their parent base, and tend to decompose with time. When these stabilized diazonium salts are used, an acid is added to their bath to a pH of 5-5.5.

For the coupling to take place at a desirable rate, without affecting the stability of the naphtholate or the diazonium salt, a specific pH must be maintained for each combination used; usually it is in the range of pH 5-7. Many times a mixture of acetic acid and sodium acetate is added to the diazonium salt solution to act as a buffer and maintain a pH of about 5. The naphtholated substrate is immersed in the diazonium salt solution at room temperature for 15-20 min., during which time the coupling reaction is completed.

Correction of unlevelness at this point of the process can be done only by stripping and re-dyeing. Since all the dyes in this group have an azo group as their chromophore they can be made colorless by reduction. The use of a proper reducing agent such as hydrosulphite ('hydro') will reduce the azo group to yield two colorless amino compounds:

$$Ar-N=N-Ar' + 2H_2 \longrightarrow Ar-NH_2 + H_2N-Ar'$$

4. After-treatment
At this stage the dyed substrate undergoes a vigorous rinsing in order to:
1) remove unreacted starting materials,
2) remove dye attached to fibers' surface, and
3) aggregate the dye molecules to improve wash-fastness and yield the final shade.

This step consists of treating the fibers at a high temperature or near the boil with a detergent, for 10 to 20 min. A slight change in shade may take place during this stage as the result of formation of aggregates (particles made of about ten to fifty dye molecules). The hydrophobic azoic dye molecules have very little attraction to cellulosic fibers. At a high temperature and with the aid of a surfactant, dye molecules move within the fiber, meet each other and form aggregates that reflect light somewhat different as compared to the light reflection of individual dye molecules. At the same time, a further improvement in wash-fastness is reached due to the increase in particle size.

Continuous Dyeing

This method is the preferred method for dyeing with naphthols, in particular when dyeing large volumes. The application range consists of: a padder, a dryer, a developing bath, and a set of washers. The application is carried out according to the following steps:

1. Naphtholation.
To minimize tailing, naphthols of low substantivity (e.g. even beta naphthol itself) are preferred. Since the substantivity of the naphtholate decreases with increasing temperature,

the padding bath is kept at temperatures as high as 80-90°C (176-194°F). Therefore, the hot dissolving method of naphthols is suitable for this application.

2. Intermediate step
The padded fabric is dryed, usually in a hot flue oven. Drying must be carried out carefully to avoid migration of the naphthol(s).

3. Coupling
The fabric is passed through the developing bath containing the diazonium salt. The pH of the developing bath is adjusted to pH 5.0-6.0, according to the nature of the coupling components. Then the fabric is passed in the air over guiding rolls (skying) to allow sufficient time for the completion of the coupling reaction.

4. After-treatment
The fabric is passed through several washers where it is first rinsed with cold water, then soaped at a high temperature, and finally rinsed with hot water.

9.3 Printing with Azoic Dyes

Azoic dyes are used in a variety of printing applications. The following are some examples of printing with azoic dyes.

Direct Printing
The fabric is impregnated with the naphthol(s), and then printed with a paste containing the diazonium salt.

It is also possible to first print with a paste containing the naphthols and then develop the color in a bath containing the diazonium salt.

Discharge Printing
The fabric is first dyed with azoic dyes in the regular manner. Then the dyed fabric is printed with a paste containing hydrosulfite and a base, and then passed through a steamer where the reduction of the azo dye in the printed areas takes place.

When a colored discharge print is desired the paste contains, in addition to hydrosulfite and sodium hydroxide, a proper vat dye. After printing, the fabric is passed through a steamer where the azoic dye in the printed areas is destroyed, and the vat dye in the printed area will become solubilized and absorbed by the cellulosic fibers (see: Application of vat dyes). The fabric is then treated with an oxidizing agent, etc., as in continuous dyeing with vat dyes.

References

1. Diazo and Azo Chemistry, H. Zollinger, Interscience, New York, 1961
2. Preston Clifford (Editor), The Dyeing of Cellulosic Fibers, 1986, S.D.C.,Dyers' Company Publications Trust, West Yorkshire, England, Chapter 8. Dyeing with Azoic Dyes, by Hans Herzog

Chapter 10

VAT DYES

The vat dyes are insoluble organic compounds that are not substantive to cellulose. Prior to dyeing they are converted to their soluble form (leuco soluble vat dye) by means of reduction in the presence of a strong base (see Fig. 10.1). In this soluble reduced form, they are substantive to cellulosic fibers, and can be applied to them. Once inside the fibers, uniformly distributed, the vat dyes are then oxidized and converted back to their original insoluble form (Fig. 10.1).

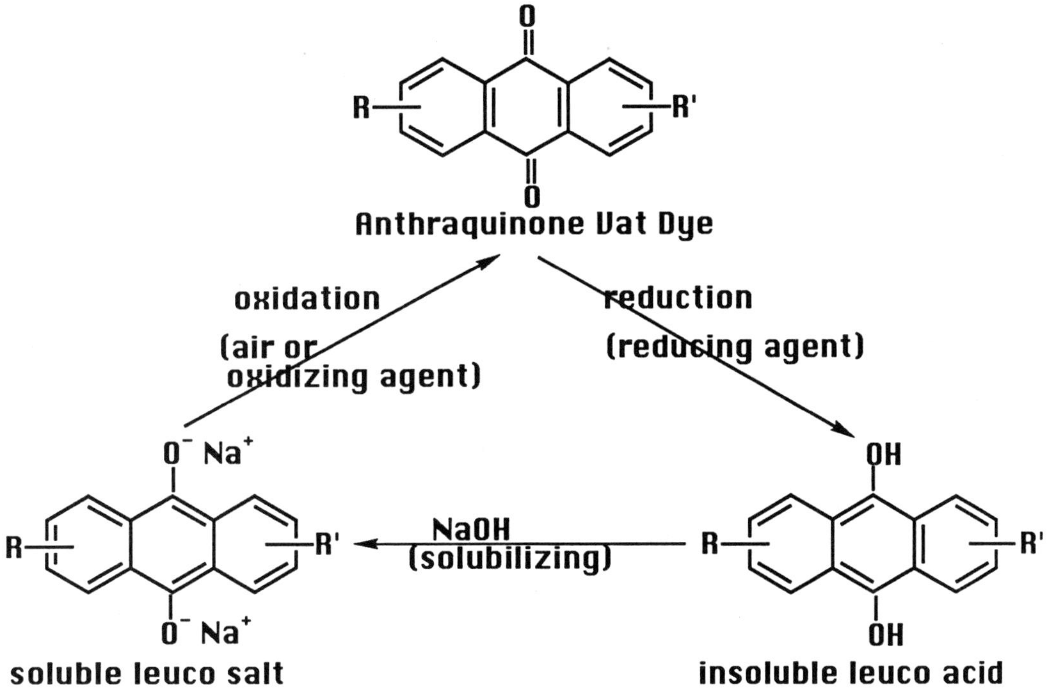

Fig. 10.1 The Changes that a Vat Dye Undergoes During its Application

The vat dyes are insoluble aromatic compounds containing two or more carbonyl groups connected through conjugated double bonds. As shown in Fig. 10.1, in the course of the reduction stage carbonyl groups of the vat dye are reduced to phenolic type groups (leuco acid vat dye) which are slightly acidic. In presence of a strong base such as sodium hydroxide the sodium salt of the reduced vat dye is formed. After the dyeing stage the dye is oxidized and remains inside the fibers in its original insoluble form. Oxidation can take place by exposing the fibers to air, or by using oxidizing agents as done in production to expedite the process.

There are two main types of vat dyes: the anthraquinone derivatives and the indigoids (indigo, thioindigo and their derivatives). The great majority of the vat dyes are the anthraquinone derivatives. Examples of the two types are:

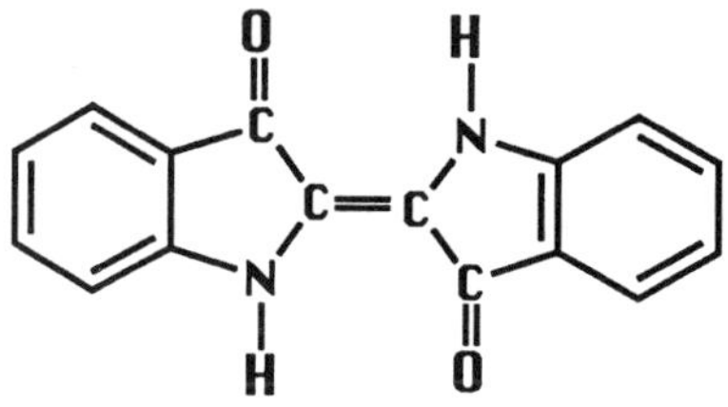

C.I. Vat Blue 1   Indigo   Dyeing Method: Special

and,

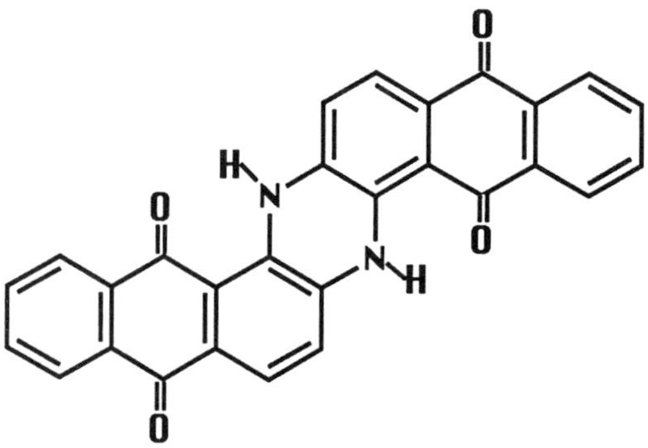

C.I. Vat Blue 4   Indanthrone   Dyeing Method: IN
C.I. Pigment Blue 60

10.1 History

Dyeing with vat dyes began with indigo and tyrian purple (dibromo indigo) which were among the oldest natural dyes known in ancient times. Indigo has a history as old as human civilization and is still being used extensively today as a vat dye (e.g. for blue denim). Textile clothes dyed with indigo have been found on Egyptian mummies believed to be at least 4000 years old. Other garments dyed with indigo were found in India and in ancient graves of the Incas in Peru and Chile. The use of indigo as a blue dye was most likely first practiced in India where it was extracted from the leaves of indigo plants (e.g. Indigofera-tinctoria). From there, growing of plants containing indigo spread gradually to other countries. In Europe, for a great many years indigo was recovered from the Woad plant which contains the dye to a much lesser extent as compared to the amount found on the Indigo-fera plants.

In 1780, German chemists showed that the dyes extracted from Indigo-fera and Woad plants were identical. A hundred years later (1879) the chemical structure of indigo was successfully elucidated by Adolf Baeyer and others. Synthetic indigo was first marketed by Badishe Anilin und Soda Fabric (BASF) in 1897, at which time the struggle between new synthetic dyes and the natural dyes reached its peak. At the outbreak of World War I, synthetic indigo replaced almost all of the natural product..

Tyrian purple was obtained from mullusks (sea snails) found at certain locations along the Mediterranean sea. Many thousands of snails were needed to produce enough of the purple of the ancients to dye a single robe, and so its use was confined to the royalty and other high ranking dignitaries who could afford it. The structure of Tyrian purple was elucidated in 1906 by Friedlander and found to be 6,6'-dibromo-indigo. This dye has no commercial value since other vat dyes with similar colors are available which posses superior properties.

10.2 Properties of Vat Dyes

Vat dyes are almost always used on cellulosic materials. They can, however, also be used on protein and nylon fibers.

The indigoids have brilliant colors with outstanding high intensities, but not satisfactory fastness properties. In particular, the indigoids are rather sensitive to sunlight. In contrast the anthraquinone derivatives exhibit overall outstanding fastness properties, but have colors of relativelyy low intensity (large amounts of dye are required for building up deep shades) and less brightness. Most of the anthraquinone vat dyes have wash-fastness ratings of 4-5, and show very good to excellent light-fastness. They also have very good fastness to bleaching, perspiration, crocking, and high temperature treatments.

The first anthraquinone derivative vat dye of a commercial value was Indanthrone synthesized by Rene Bohn (from BASF) in 1901 (C.I. Vat Blue 4, see above). Since then hundreds of new anthraquinone vat dyes, have been synthesized. Vat dyes are now available in a wide range of shades with weakness in bright reds.

The reduced form of Indigo is colorless and has been called leuco-indigo (leuco in Greek is white), and the term 'leuco' is now used for the reduced form of any vat dye. This change of color, not necessarily to white, is typical to all vat dyes. It is due to the change in the structure of their chromogen that takes place during reduction. In the course of reduction the carbonyl groups that act as electron acceptors in the chromogen are changed to phenolic type groups which act as electron donors, resulting in a dramatic change in color. A color change is observed once again when the leuco dye is treated with sodium hydroxide. The color of the sodium salt of a leuco vat dye thus obtained, differs from both the colors of the leuco acid dye and the vat dye itself.

10.3 The Application of Vat Dyes

The conventional exhaust method of vat dyes consists of the following four major steps:
Reduction ('vatting')
Dyeing
Oxidation
Soap at the boil

1. Reduction (vatting). At this stage the vat dye is converted into its soluble form. The dye is first mixed with the proper amount of sodium hydroxide. Then the reducing agent is added, and the temperature is raised to the recommended temperature for the reduction reaction to occur as shown in Table 10.1. Ease of reduction may vary greatly from one vat dye to another, and the amounts of reducing agent and base needed for specific dyes are shown in Table 10.1. The reduction time also varies greatly, from between two to 30 min., depending on the specific dyes used.

The reducing agent most commonly used is sodium dithionite: $Na_2S_2O_4$, incorrectly called 'hydrosulphite', and also referred to as 'hydro'. During the basic reduction, hydrosulphite is oxidized to sodium sulphite and sodium sulfate:

$Na_2S_2O_4 + 2NaOH + 2O \longrightarrow Na_2SO_3 + Na_2SO_4 + H_2O$

About half of the sodium hydroxide used is consumed in the reduction reaction, and the other half reacts with the leuco acid to form its salt. This reaction also takes place with oxygen in the water and the atmosphere causing some loss of the reducing agent and the base during application.

Hydrosulphite will oxidize on exposure to air in presence of moisture or in solutions, and may even ignite when kept in storage under moist conditions. It is also very unstable at high temperatures, and therefore should be dissolved in cold water just before adding it to the dye-bath.

Another reducing agent not often used for vat dyes is thiourea-dioxide: $O_2S-C(NH_2)_2$. In presence of NaOH, thiourea decomposes to yield the sodium salt of sulfoxylic acid which is a very strong reducing agent:

$$O_2S-C(NH_2)_2 + 2NaOH \longrightarrow (NH_2)_2CO + Na_2SO_2 + H_2O$$

Thiourea-dioxide is much more expensive than hydrosulphite. According to thiourea-dioxide manufacturers, a much smaller amount of this chemical (1/5 to 1/10) as compared to the amount of 'hydrosulphite' is needed to reduce the same amount of a vat dye. Therefore, thiourea-dioxide is easier to handle, and far less storage place is required. Thiourea-dioxide is also significantly more stable than hydrosulphite, and is not affected by moisture.

The use of soft water throughout the dyeing stage is a must with vat dyes, since the soluble leuco salts will form insoluble salts with calcium or magnesium ions as well as with transition metals. Therefore, in addition to using soft water, it is common to add sequestering agents such as EDTA to the dye-bath.

Certain vat dyes may undergo over-reduction if reduction procedures are not followed carefully. In over-reduction the leuco vat dye is further reduced into a structure that will not oxidize back under the normal conditions used in the application. Factors leading to over-reduction are: use of large excess of reducing agents, insufficient amount of NaOH in dye-bath (pH should be above 7), and dyeing at higher temperatures than those recommended. The leuco acid of the vat dye tends to be reduced further more so than the sodium salt of the leuco acid. Therefore, in application procedures it is recommended to add the NaOH first, before the reducing agent.

2. Dyeing. The scoured material is inserted in the dye-bath and the temperature is gradually raised to the dyeing temperature (27°C to 60°C(80°F to 140°F), depending on the type of dyes used). Dyeing continues for the proper amount of time (Table 10.1.) and salt may be added to assist in exhaustion. Since atmospheric oxygen reacts with the reducing agent as well as with the reduced vat dyes, additional amounts of the reducing agent and the base are

added during the dyeing stage. The dye-bath is checked occasionally to insure that the pH of the bath is sufficiently basic (phenol-phthalein paper should turn red), and that a sufficient amount of the reducing agent is present in the dye-bath (yellow vat testing paper should turn blue).

3. Oxidation. Before oxidation, the material is rinsed to remove residues of sodium hydroxide and reducing agent. Potassium dichromate used to be a common oxidizing agent for vat dyes, but now seldom used because of ecological reasons. Common oxidizing agents used today are hydrogen peroxide and sodium perborate. When using hydrogen-peroxide, high concentrations of alkali must be avoided to prevent damage to fibers. Also, certain Indanthrone vat dyes tend to over-oxidize at high pH, and change color. However, the original color of the dye can be restored by reduction.

4. Soap at the boil. This step consists of treating the dyed material with a soap or surfactant solution at or near the boil for 10-20 min. Soap at the boil increases the wash-fastness of the dyes and yields their final shade. The slight change in shade that takes place during this stage is the result of the formation of aggregates which reflect light some what different than the individual dye molecules. The hydrophobic vat dye molecules have very little attraction to cellulosic fibers. Therefore, during this step, high temperatures and the aid of a surfactant cause the dye molecules to move within the fiber and meet each other to form aggregates. In this form the dye acquires even better wash-fastness.

The dyeing procedure is then completed by rinsing with hot and cold water.

Table 10.1 Application Conditions for Various Types of Vat Dyes

(approximate values for dyeing medium shades at a liquor ratio of 1:10)

TYPE	IN	IW	IK
	C.I. method 1	C.I. method 2	C.I. method 3
Reduction Temp	140°F	120-130°F	100-120°F
Dyeing Temp	140°F	110-120°F	70-80°F
NaOH	6-9 g/l	3-5 g.l	2.5-4 g.l
Hydrosulphite	6-9 g/l	4-6 g/l	3-4.5 g/l
Salt	none	7.5-15 g/l	10-20 g/l

N = normal procedure

W = warm temperature

K = cold temperature (Cold in German is Kalt)

10.4 Classification of Vat Dyes

Vat dyes differ from each other in their detailed optimum application conditions. While certain vat dyes are easy to reduce, others require more vigorous conditions (higher temperatures and larger quantities of 'hydro' and NaOH). Accordingly, the vat dyes are divided into the following groups:

IN, IK, IW, and IN special. Of the different vat dyes used nowadays the majority are of the IN type. The IW are used to a lesser extent, where as the IK vat dyes are seldom used today.

Examples of IN vat dyes are C.I. Vat Black 25, and C.I. Vat Green 1:

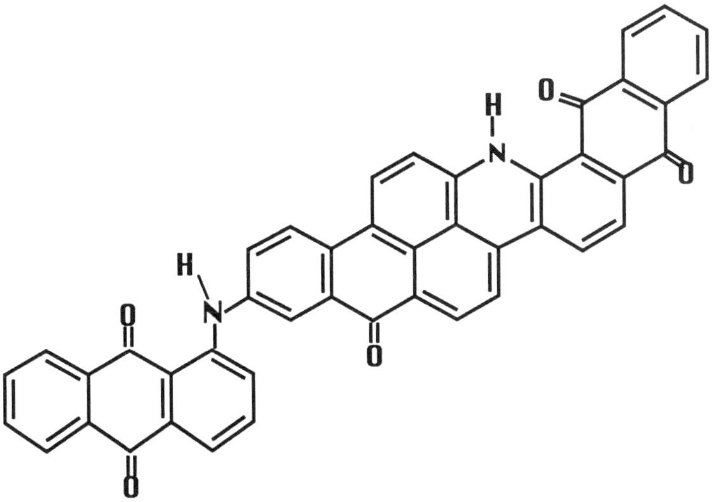

C.I. Vat Black 25   **Dyeing Method IN**

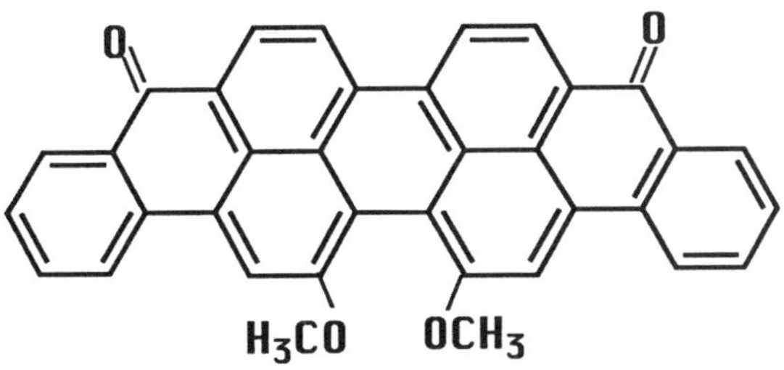

C.I. Vat Green 1   **Dyeing Method: IN (also IW, and IK)**

IN vat dyes are usually of relatively very large molecular size, and in their soluble forms exhibit high substantivity. As such they show a low tendency to dissolve in water. Therefore, higher temperatures and large amounts of 'hydro' and NaOH are required for their solubilizing process (see Table 10.1). Since the base and the reducing agent (and its oxidized products) are strong electrolytes, and since the soluble form of the dye has a high substantivity, IN vat dyes tend to strike fast. Therefore, salts are not used during their application to promote exhaustion.

In contrast the IK vat dyes are usually of a relatively small molecular size, and are easy to dissolve. They require lower temperatures and small amounts of the reducing agent and NaOH for their solubilizing process. In their reduced soluble form, IK vat dyes possess a low substantivity, and therefore require large amounts of electrolytes for exhaustion. Since their dye-bath contains relatively small amounts of electrolytes, large amounts of salts are used for exhaustion (Table 10.1).

Pigment Dyeing Method

In order to reduce problems associated with a fast strike, pigment application methods have been developed. In these methods the substrate is first treated with a dispersion of the vat dye in its insoluble pigment form. The dye at this stage is evenly distributed on the fibers' surface. Next the base and the reducing agent are added, and the solubilized dye is absorbed by the fibers. Dyeing is then continued in the regular manner.

The dyes used in this method must have an extremely fine particle size. Pigmentation is carried out by circulating the dye dispersion through the fibers and/or by circulating the substrate through the dye-bath. High temperatures are usually used to insure an even application of the pigment. Pigmentation can also be achieved by padding a fabric with a dispersion of vat dyes. After padding, the fabric is placed in a bath containing the reducing agent and the base, and reduction takes place as in exhaust dyeing.

10.5 Application of Vat Dyes by Continuous Methods

The Pigmentation Process   (dye-pad / dry / chemical-pad / steam)
This method is known to produce very good results, with very good reproducibility in both light and dark shades and on different types of fabrics. In this process, the vat dyes in their normal insoluble state (pigment form) are applied to the fabric by a padder. After padding, the fabric is dried, passed through a second padder containing the reducing agent and sodium hydroxide, and then introduced into a steamer. In the steamer the dyes are reduced

and become water soluble. In this form, they diffuse into the fibers which are now readily accessible. Then the fabric is passed through a set of washers where the remaining steps of the dyeing process take place: rinsing, oxidizing, soaping at the boil, and final rinsing.

The following is a description of the steps involved in continuous dyeing with vat dyes by the pigmentation method.

Preparation. Proper preparation is more critical with continuous dyeing than with exhaust dyeing. The fabric must be uniformly prepared, and have good absorbency. Also, the moisture content of the fabric before entering the padder must be as uniform as possible, and preferably within a few percent of the moisture regain of cotton.

Entering the padder, the fabric should not be too wet nor too dry ('bone dry'). After preparation, the fabric is in a wet state, and it will be far more economical to start dyeing without first drying (wet-on-wet application). However, the problem with this approach is that it is difficult to control a uniform moisture content of a wet or partially wet fabric. In addition, it is not recommended to pad a wet fabric since the exchange of liquids decreases the dye concentration in the padding-bath. On the other hand the fabric should not be over dryed either since it becomes difficult to wet it properly in the very short time that the fabric is in contact with the dye-bath. When water is completely removed from cotton, the cellulosic chains in the amorphous regions get closer to each other, causing additional hydrogen bonds to be formed between them, making it more difficult for water to re-enter the fiber. A small amount of water is sufficient to swell the amorphous regions enough so that the fabric will readily absorb the dye solution. Therefore, the fabric with its normal moisture regain content is in the proper condition for padding.

Dye-pad. The dye-bath contains the vat dyes, an antimigrant and a wetting agent. The vat dyes are usually of the IN type but other types are also used. Dyes in liquid form are preferred, since they contain lower quantities of dispersing agents which may enhance dye-migration during the drying stage that follows. Typical antimigrants are thickening agents such as sodium alginate and acrylic acid copolymers. These chemicals combine with dye-particles by weak attractions and prevent them from moving along with the evaporating water. A surfactant which is more hydrophilic than hydrophobic can serve as a wetting agent.

Padding is performed at a high pressure to obtain as low wet pick-up as possible. The larger the amount of water to be removed the more likely that water will evaporate unevenly. Therefore the squeeze rolls are adjusted to attain a wet pick-up of approximately 60-70%.

Pre-dry. Infra-red ovens are used to remove the first half of the water which is more liable to cause dye-migration. In these non-contact drying ovens the water is removed more uniformly than in other conventional drying ovens.

Dry. Drying is completed in a hot-flue oven, or on heated cans.

Chemical-pad. The chemical-pad contains the reducing agent ('hydro') and sodium hydroxide that are needed for solubilizing the vat dyes. A high wet pick-up in the range of 90 to 100% is preferred to prevent the presence of air in fabric which may interfere with the reduction process.

Steam. In the steamer, the dyes are reduced and solubilized, and then diffuse into the cellulosic fibers. Saturated or super-saturated steam at a temperature of 102-105°C (215-220°F) is used, and the steaming time is between 30 to 60 sec. Dry steam at higher temperatures (super-heated steam) may dry the fabric and cause a reduction in the yield of dyeing. This is because a certain amount of moisture on fabric is needed for the proper reduction and diffusion of the dye. The steamer must operate under an air-free atmosphere. Oxygen in the steamer will cause a pre-mature oxidation of dye, resulting in loss of yield.

After-treatment. The fabric is passed through a set of about ten wash boxes to oxidize, and remove all dyeing assistants and dyes on surface of fibers. First the fabric is rinsed with cold water to remove unreacted reducing agent and alkali that may interfere with oxidation. Next oxidation takes place, usually with hydrogen peroxide at a neutral pH. The dyeing process is completed by soaping at the boil, and rinsing with warm and cold water.

10.6 Soluble Vat Dyes

The soluble vat dyes are actually a group of dyes by themselves. They are vat dyes already in the reduced soluble form, and have the unique feature that they will not oxidize by merely exposing them to the air. The soluble vat dyes are sulfuric acid esters of the leuco vat dyes. Their oxidation requires the use of a proper oxidizing agent (e.g. sodium nitrite) in a strongly acidic solution. In presence of the acid and at the proper temperature, the ester will hydrolyze to yield the leuco acid of the vat dye which can then be oxidized even by air, as are regular leuco vats, to its insoluble form. The first soluble vat dye was made from indigo as shown in Fig. 10.2:

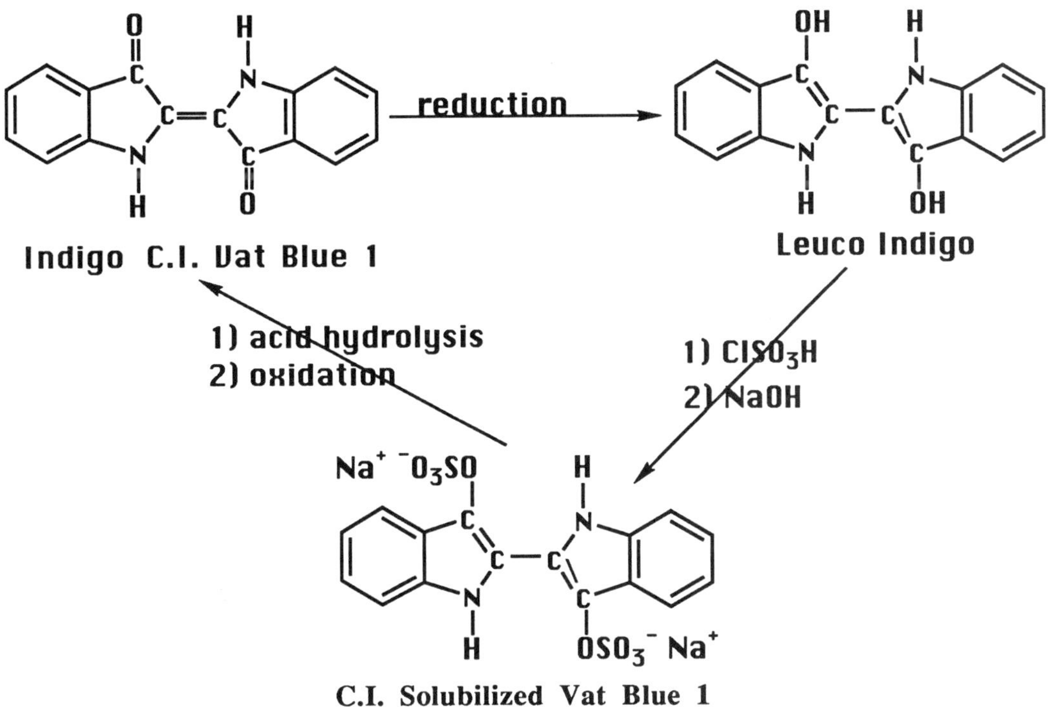

Fig. 10.2 Synthesis and Hydrolysis of Soluble Vat Indigo

Soluble vat dyes offer the following advantages:
1. The reduction step is eliminated and there is no need for careful control of the dye-bath during application.
2. The application is more easy to follow since there are not large amounts of electrolytes (NaOH, 'hydro', etc.) in the dye-bath that may cause a fast- strike.
3. The soluble vat dyes can be applied to protein fibers that are sensitive to a basic pH.

The main disadvantages of the soluble vat dyes are their high cost and their limited shelf life. When left for long periods of time in presence of moisture, they will hydrolyze and convert back to the regular vat dye.

Soluble vat dyes have been rarely used in the last decade. However, with the recently growing interest in cotton/wool blends there might be a new interest in the soluble vat dyes since they can be used for union dyeing of these blends.

References

1. The Application of Vat Dyes, AATCC Monograph No. 2, Lowell, Massachusetts, 1953.
2. Dyeing with Indanthren Dyes, Technical Information, BASF, 1988.
3. John H. Polevy, et al., Influence of Vat Dye Particle Size on Color Yield and Industrial Wash-fastness, Book of Papers, AATCC 1990 International Conference &Exhibition, p. 12.

Chapter 11.

SULFUR DYES

Sulfur dyes are insoluble in water and their application to cellulosic fibers resembles that of the vat dyes in principle. Sulfur dyes are also made water-soluble by a treatment with an alkali solution of a reducing agent. In their soluble form sulfur dyes are substantive to cellulose, and are applied as such. After the dyeing steps they are converted back into an insoluble form (but not necessarily to their original insoluble form), by a treatment with an oxidizing agent.

Sulfur dyes are easier to reduce and solubilize as compared to vat dyes. A mild reducing treatment, such as sodium polysulfide, and a small amount of sodium carbonate are used to solubilize sulfur dyes. Also, reduced sulfur dyes will not oxidize easily by merely exposing them to the air, as do leuco vat dyes.

Sulfur dyes are made by reacting aromatic amines or nitro-benzene derivatives with sulfur and/or sodium sulfide at high temperatures in sealed containers. This type of a reaction yields a mixture of dyes with complex structures. These dyes contain sulfur in their structure, hence the name 'sulfur' dyes. No attempt has been made to isolate the individual dyes in the mixture because the procedure is extremely difficult. Therefore the reaction product is used 'as is'.

11.1 The Chemical Nature of Sulfur Dyes

In general sulfur dyes have a high molecular weight and a complex structure. Information on some parts of individual sulfur dyes is available. The following are typical parts of chromogens found in sulfur dyes [1]:

The reducible sites are in the form of disulfides or polysulfides and upon reduction in presence of a base the soluble mercapto salt is formed:

$$\text{DYE-S-S-DYE} \xrightarrow{Na_2S} 2 \text{ DYE-S}^-Na^+$$

$$\text{DYE-S}_n\text{-S-H} \xrightarrow{Na_2S} \text{DYE-S}^-Na^+$$

Upon oxidation, disulfide bonds can be formed between two dye molecules:

$$\text{DYE-S}^-Na^+ \xrightarrow{NaBrO_3} \text{DYE-S-S-DYE}$$

The resulting increase in molecular size further increases the wash-fastness of the sulfur dyes.

11.2 General Properties of Sulfur Dyes

Sulfur dyes are extensively used since they provide an attractive combination of a relatively easy way to dye cellulosics with good-to- excellent wash and light-fastness at a low cost. This is the result of their inexpensive preparation, starting from simple raw materials and using the reaction products as obtained without further purifications. Their method of application is easy to follow, without the complications associated with vat dyes, such as the possibility of 'over-reduction' and sensitivity to variations in temperature.

The main disadvantages of sulfur dyes is their color limitations in bright colors, and their poor light-fastness in pastel shades. Therefore they are used mainly for dyeing cellulosics in black, brown, navy blue, or olive, in medium to dark shades. Some yellows and blues are available, but there is only one red and one green dye.

Note that the red and the green dyes have unexpected brightness for a sulfur dye, and are not true sulfur dyes. They are vat dyes that can be applied together with sulfur dyes under similar conditions. Even though they are more expensive than sulfur dyes they are offered to be used together with them for shading purposes, to extend the spectrum of colors that can be obtained with sulfur dyes.

Pottasium bichromate in presence of acetic acid used to be the preferred oxidation system. However, because of strict State and Federal regulations, the allowed amounts of chrome compounds in the effluent are extremely small. Thus, this method is seldom used. Currently, oxidation of sulfur dyes is performed with mild reducing agents, such as sodium bromate or potassium iodate. Stronger oxidizing agents, such as hydrogen peroxide, should be used carefully; otherwise they may cause loss of dye by over-oxidizing the sulfur dye into a new soluble form. For that reason sulfur dyes are sensitive to chlorine or peroxide bleaching.

Sulfur dyes on material stored at higher than normal room temperature and in presence of moisture tend to oxidize to form strong sulfur acids. These acids will then cause tendering of cellulosic fibers. To prevent tendering, after dyeing, the final rinse is carried out with a mild alkali solution.

11.3 The Application of Sulfur Dyes

Sulfur dyes are applied by exhaust [2, 3, 4] or continuous methods [3, 4, 5]. In exhaust dyeing, the soluble sulfur dye is applied in a manner similar to that of the direct dye application. Continuous dyeing by the pad/steam method is the most widely-used method for applying sulfur dyes.

Currently, sulfur dyes are supplied mainly in their pre-reduced state, in liquid form, as 'Soluble Sulfur Dyes'. The insoluble solid form of sulfur dyes is somewhat difficult to solubilize, and undissolved particles left in the dye-bath may interfere with the dyeing. The Soluble Sulfur Dyes (e.g. Sodyesul Liquids by Sandoz Chem. Corp.) are supplied as uniform concentrated solutions and are much easier to handle. They can be accurately measured and then diluted by adding them directly to the dye-bath. Though the sulfur dyes are already in their soluble form, additional amounts of the reducing agent (sodium polysulfide) are added to insure a reducing environment within the dye-bath.

During the entire dyeing stage, the pH of the dye-bath must be kept on the alkaline side. An added acid will react with sulfides in the dye-bath to release hydrogen sulfide which is highly toxic. Therefore, after dyeing and before oxidizing, the dyed fibers are rinsed well with warm and hot water. This is done to remove unfixed dye on the surface of fibers, and to remove unreacted sulfides. If left on fibers, the toxic hydrogen sulfide will be released in the oxidation step that follows which is carried out under acidic conditions. There is no risk of losing the soluble sulfur dye during the extensive rinsing since it is highly substantive to the cellulosic fibers.

Soluble forms of sulfur dyes may combine with heavy metals to form insoluble salts, and therefore sequestering agents, such as EDTA are usually added to the dye-bath.

Exhaust Dyeing With Sulfur Dyes

Soluble sulfur dyes behave like direct dyes with high substantivity. They tend to exhaust at lower temperatures and at a higher rate. Salt is added carefully at the lower temperatures. Then, the temperature is gradually raised to the maximum dyeing temperature to complete exhaustion.

A typical batch dyeing is carried out as follows:
The dye-bath is prepared by diluting the Soluble Sulfur Dye with the proper amount of water. A small amount of sodium polysulfide (e.g. 1-2% o.w.f.) and a similar amount of a sequestering agent are added, and the fibers are introduced into the dye-bath at 50°C (122°F). Sodium chloride (20-30% o.w.f.) is added portion-wise (e.g. 1/8, 1/8, 1/4, and 1/2 of the total amount) over a period of 20 min., while the temperature is gradually raised to the maximum dyeing temperature, usually in the range of 60-82°C (140-180°F). Dyeing continues at this temperature for about 30 min. The fibers are then rinsed well with warm and hot water, oxidized, rinsed, soaped at 82°C (180°F), rinsed with a mild alkali solution, and dried.

Continuous Dyeing With Sulfur Dyes

Sulfur dyes are applied continuously by the pad/steam method which consists of the following steps: pad/steam/rinse/oxidize/rinse.

Pad

The padding bath contains the soluble sulfur dyes, sodium polysulfide, a wetting agent, and a sequestering agent. Since the soluble sulfur dyes are highly substantive to cellulose, precautions must be taken to control tailing [5]. If no adjustments are made and the feeding solution has the same content as the initial content of the dye-bath, tailing will take place until the system reaches equilibrium. In this method, the trough should be as small as possible to minimize the initial amount of shade off-fabric. To avoid tailing from the start, the initial dye-bath is diluted by approximately 10-15%. This requires information regarding concentration of dyes and the specific dyes used.

A wet pick-up in the range of 80% is preferred to minimize the amount of air left in fabric.

Steam

The steamer is filled with steam to obtain an oxygen-free atmosphere. Steaming is performed at 102-104°C (216-219°F) for one minute.

Rinse

The fabric is passed through a few wash boxes and thoroughly rinsed with warm and hot water.

Oxidize

Currently the most widely-used method is oxidation with sodium bromate and a vanadium compound as a catalyst.

Rinse

The fabric is first rinsed with hot water, and then with cold water containing a small amount of sodium carbonate.

References

1. W. Zerweck, et al., Angew. Chem., 60A <u>1948</u>, 141
2. Tobin H, American Dyestuff Reporter, 68, p.26, 1979
3. Leon Tigler, Textile Chemist & Colourists, Vol.12, No. 6, p. 43, 1980
4. Sulfur Dyes, Martin Marietta Chemicals, Sodyeco Division, 2/80-LT
5. Tom Burns, Efficient Pad Steam of Sulfur Colors, Sandoz Chem. Corp., December 1, 1987.

Chapter 12

REACTIVE DYES

The reactive dyes are water-soluble anionic dyes which react with hydroxyl groups of cellulose to become covalently bonded to the cellulosic fiber. An example of a reactive dye is C. I. Reactive Blue 5 :

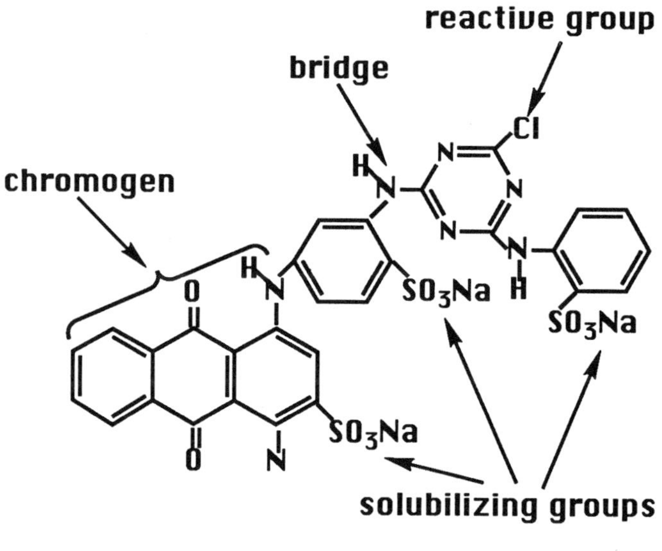

C.I. Reactive Blue 5   (monochlorotriazinyl)

Fig. 12.1 The components of a Reactive Dye

As indicated in Fig. 12.1, the molecular structure of a reactive dye consists of: a chromogen(s), a solubilizing group(s), a bridge, and a reactive group(s). The conjugated double bonds system of the chromogen is discontinued at the bridge (-NH-). In this way the reactive group itself or any change that takes place at the site of the reactive group will not have any effect on the color. The chemical reaction between a reactive dye and a cellulosic fiber takes place in the presence of a base and can be summarized as follows:

$$\text{DYE-Cl} + \text{H-O-cellulose} \xrightarrow{\text{base (-HCl)}} \text{DYE-O-cellulose} + \text{salt}$$

The covalent bond thus formed provides very good wash-fastness, and is far stronger than the weak hydrogen bonds of a direct dye with cellulose.

Reactive dyes react in a similar way with amino groups as well as with the hydroxyl groups of water. The reaction with amino groups enables their use on protein fibers and nylons.

In the reaction with water the reactive group is hydrolyzed and the dye looses its ability to react with hydroxyl groups of cotton or amino groups of polyamides:

$$\text{DYE-Cl} + \text{H-OH} \xrightarrow{\text{base (-HCl)}} \text{DYE-OH} + \text{salt}$$

Even though the reactive dyes can react with both cellulosic fibers and water, the reaction occurs mainly with the fibers for two main reasons: (1) the rate of the chemical reaction between a reactive dye and cellulose is much faster than its reaction with water ($\approx 100:1$), (2) the probability of a diffused dye molecule reacting with an OH group on cellulose is much lager than its probability to react with an OH group of water [1]. Therefore, in the dyeing of cellulose with reactive dyes most of the dye reacts with the fibers, and only a small amount of it reacts with the water of the dye-bath. The amount of unfixed dye, however, which is composed of hydrolyzed dye and unreacted dye, could be as high as 20-30% of the total amount of dye used.

It is interesting to note that reactive dyes have become commercially available only since 1956, a hundred years after the introduction of the first synthetic dye Mauve, by Perkin. Until 1956, many attempts were made to bond dyes to cellulose by covalent linkages with the hope of achieving superior wash-fastness. However, in all these trials the reaction conditions were too severe and caused excessive reduction in the strength of the fibers. The first available reactive dyes for cellulosic fibers were the Procion dyes of Imperial Chemical Industries (I.C.I.). These dyes could be bonded to cellulosic fibers under mild conditions without noticeable damage.

12.1 The Chemical Nature of Reactive Dyes

The Procion Reactive Dyes

The starting material for the synthesis of Procion dyes is Cyanuric-chloride (Trichloro-1,3,5 triazine):

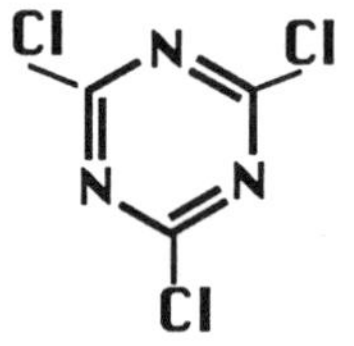

The chlorines in this molecule react readily with hydroxyl groups, amino groups, and water, through nucleophilic substitution reactions. The reactivity of the chlorines decreases with decreasing the number of chlorine atoms on the ring. Thus, the first chlorine within trichlorotriazine will react by just mixing the compound with alcohols, amines, or water, at room temperature. For the second chlorine to react with these compounds, the presence of a weak base (e.g. $NaHCO_3$) is required. When only one chlorine is left on the triazine ring, the chemical reaction will take place only at high temperatures and in the presence of a stronger base (e.g. Na_2CO_3). In making a Procion dye, the first chlorine is used to bond a chromogen to the triazine ring through an amino bridge:

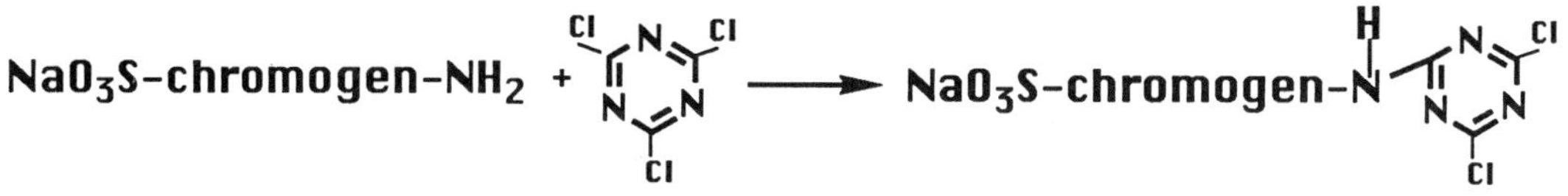

NaO_3S-chromogen-NH_2 + [triazine with 3 Cl] ⟶ NaO_3S-chromogen-N(H)-[triazine with 2 Cl]

The reactive dye thus obtained contains two reactive chlorines and will react with cellulose in presence of sodium carbonate at a temperature as low as room temperature. If a less reactive dye is desired, the second chlorine of the triazine is reacted, for example with aniline, to yield a dye with only one reactive chlorine (e.g. C.I. Reactive Blue 5, Fig. 12.1). This dye is classified as a Procion H (H-hot) type and will react with cellulose in presence of sodium carbonate at temperatures in the range of 77-82°C (170°-180°F).

Vinyl Sulfone Reactive Dyes

Another group of reactive dyes contain the sulfato-ethyl-sulfone group ($-SO_2-CH_2-CH_2-OSO_3H$) as their reactive group (e.g. the Ramazol dyes of Hoechst). In the presence of a base this group undergoes an elimination reaction to form a vinyl-sulfone group ($-SO_2-CH=CH_2$) which then combines with cellulose through an addition reaction:

1. $DYE-SO_2-CH_2-CH_2-OSO_3H + NaOH \longrightarrow DYE-SO_2-CH=CH_2 + NaHSO_4$

2. $DYE-SO_2-CH=CH_2 + H-O-cell. \longrightarrow DYE-SO_2-CH_2-CH_2-O-cell.$

The degree of reactivity of the sulphato-ethyl-sulfone group is between that of a monochloro-triazine and a dichloro-triazine.

Since reactive dyes from different dye manufacturers may differ greatly in their chemical reactivity, they should not be mixed in the same application. Even when using reactive dyes of the same company, the selected dyes should belong to the same subgroup, or mixed as recommended by the manufacturer.

12.2 The Properties of Reactive Dyes

The reactive dyes provide the following attractive advantages:

a. available in a complete range of colors including very bright colors

b. very good wash-fastness

c. good to very good light-fastness

d. high flexibility in the choice of method of application

e. easy to obtain level dyeing

f. readily soluble in water.

However, the cost of using reactive dyes is high. This is due to the prices of the reactive dyes, the loss of substantial amounts of dye during application, and the excessive time required for the dyeing process. Another disadvantage of the reactive dyes is that many of them are sensitive to oxidation, in particular to the effect of chlorine. Also, because of the relatively high loss of dyes during application, and the effect of adding the base on the rate of exhaustion, reproducible results are more difficult to obtain.

12.3 The Application of Reactive Dyes By Exhaust Methods

Applications of reactive dyes by batch dyeing consist of three stages:

1. exhaust dyeing,
2. fixation, and
3. after-scouring.

In this procedure the dyes are first introduced into the fibers, using an electrolyte to assist in exhaustion. When most of the dye has penetrated the fibers properly, the base is added, and the dye-bath temperature is brought to the fixation temperature, at which time the chemical reaction between the dye and the fibers occurs. The third step in the process involves an extensive scouring procedure aiming at removing unbonded dyes that may be attached to the fibers by weak attractions

When dyeing with reactive dyes the cellulosic fibers must be free of any chemicals that may react with the dyes. Woven fabrics, for example, must be thoroughly desized, as reactive dyes will react with hydroxyl groups of sizing materials, such as starches or polyvinyl alcohol, the same way they react with cellulose. The fibers should also be free of any residual bases from previous treatments such as caustic scouring, mercerizing, or stripping.

The following is a typical procedure for exhaust dyeing at a high temperature:
The dye-bath temperature is set at 50°C (122°F), and the pre-dissolved reactive dyes are added. Then the substrate is placed into the dye-bath and the temperature is raised at 1°C per min. to 80°C (176°F) (about 30 min.). During this time the salt (e.g. 100% o.w.f., depending on liquor ratio and depth of shade) is added portion-wise. Dyeing continues at 80°C (176°F) for an additional 15 min., after-which sodium carbonate is added over 15 min. The dyeing machine is run at 80°C (176°F) for 45-75 min., during which time fixation is completed. The dyed material is rinsed several times with warm water, to remove the salts and the base, and then soaped at the boil, to remove unfixed dyes. Finally, the material is given two more rinses, first with warm water followed by cold water, and then dried.

Dye Substantivity

One of the main problems associated with the application of reactive dyes is the removal of the unbonded dye. If not properly removed, this portion of the dye, will adhere to the fibers like a direct dye and will come-off in subsequent home launderings. Therefore, the reactive dyes are made by dye manufacturers to have a relatively low substantivity. It will then be easier to remove unbonded dyes of low substantivity by proper rinsing and soaping.

The low substantivity, which is achieved in part by increasing the solubility of the reactive dye (dye molecules contain a relatively large number of sulfonic acid groups), results in very good leveling properties. They have a high rate of diffusion, and are easy to migrate. Also, the preparation of the dye-bath is easier since the dyes are readily soluble.

Salt Additions

The low substantivity of the reactive dyes leads to the necessity of using very large amounts of salt for exhaustion. The amount of salt needed can be over four times the amount used with direct dyes, when applied under similar conditions (when dyeing with similar amounts of dyes at the same liquor ratio). Since the required amount of salt depends on the liquor ratio, it should be expressed in weight of salt per volume of dye-bath (e.g. grams/liter). The required amount of salt increases with increasing amount of dyes used, and is in the range of 20-100 g/l, and more. Some dyeing procedures recommend to add all the salt at the beginning before the addition of the dyes. Other procedures recommend to add the salt after the dyes, before raising the temperature to the dyeing temperature. Still other procedures recommend adding the salt and the base before adding the dyes.

Liquor Ratio

Because of their low substantivity and very good solubility, a considerable amount of the dyes is left in the dye-bath. Therefore, for a high efficiency in the utilization of the dyes, dyeing at lower liquor ratios is preferred. Using smaller amounts of water will also help in avoiding additional loss of dye due to hydrolysis.

Alkali Additions

Besides sodium carbonate, other bases are used for fixation. Among these are: sodium hydroxide, trisodium phosphate, sodium silicate, and sodium bicarbonate. Sometimes, combinations of the above, such as mixtures of sodium hydroxide and sodium silicate, or sodium hydroxide and sodium carbonate, are used.

At the time of adding the base, a substantial amount of the dye may still be present in the dye-bath. The addition of the base can promote further exhaustion due to two reasons: first, the chemical bonding of the dye to the fiber drives the dynamic equilibrium of the dye molecules' movements towards exhaustion. Therefore, the faster the chemical reaction occurs, the faster the rate of exhaustion. Second, the ionic nature of the base further enhances the exhaustion of the dyes. Therefore, to avoid unlevel dyeing, the base is added portion-wise over a proper length of time, and not too early.

A significant increase in exhaustion during the base addition is more often observed with the vinyl sulfone reactive dyes. The part of the dye that is still in the dye-bath will lose its sulfato group through the elimination reaction to yield the vinyl form of the dye which is much less water soluble and will exhaust much faster.

Equipment is available today that allows continuous addition of base solutions in a fully-controlled manner, over a specific period of time, as desired.

Dyeing Temperature

The rate of reaction depends not only on the reactivity of the reactive dyes, but also on the temperature, strength of the base added, and fixation time. Accordingly reactive dyes can be applied over a large range of temperatures and conditions as shown in table 12.1.

Table 12.1 Examples of possible fixation conditions for reactive dyes

Method of Application	Base	Fixation Temperature	Fixation Time
Cold Pad/Batch	NaOH or Na_2CO_3	Around room temperature	Several hours
Exhaust dyeing	Na_2CO_3	60°C (140°F)	30-60 min.
Exhaust dyeing	$NaHCO_3$ or Na_2CO_3	82°C (180°F)	30-60 min.
Pad/steam	Na_2CO_3	100°C (212°F)	a few min.
Pad/dry/cure	Na_2CO_3	150°C (302°F)	1-3 min.

12.4 Cold Pad-Batch Method

This is a semi-continuous method. In the application, the fabric is padded with a solution containing the dyes and the base, and then wound onto a roll. The rolls are covered with plastic sheets to prevent evaporation of water, and left at room temperature, from two hours to overnight, depending on the rate of fixation. The rolls are rotated once in a while, in particular when the liquor pick-up in the padding is high, in order to prevent settling of the dye solution to the bottom. The unbonded dye is then washed off by rinsing with cold and warm water, soap at the boil, etc. The wash-off can be done on conventional available equipment such as a continuous wash-range or equipment used in exhaust dyeing.

A feeding device is used to mix the dye solution with the base solution immediately before adding them to the padding bath. Since the volume of the padding bath is made as small as possible (to avoid tailing), and the fabric is padded at a rate of at least 60 yards per minute, the dyes-base mix is added continuously using a metering system.

In the cold pad-batch method, dyes with high reactivity are preferred since fixation takes place at room temperature. Batching time should be long enough to complete fixation for the entire roll. Otherwise the part of the roll that was padded last will have less fixed dye on it.

The cold pad-batch method provides a relatively simple and inexpensive way of applying reactive dyes at a high rate to fabrics made of cotton or rayon and their blends. The equipment can be used for woven fabrics or knits with different constructions, and for short or long production lots. Impressively long lists of advantages obtained by this method appear in dye manufacturers' publications. Among the listed advantages are the low initial investment in equipment, and the small space requirements. Also, in addition to savings in chemicals and water as compared to exhaust dyeing methods, there is a substantial reduction in energy consumption since dyeing and fixation take place at room temperature. Very good shade reproducibility with high color yields are also claimed.

However, to avoid off-shade dyeing, which are difficult to correct in padding applications, the process has to be fully controlled (e.g. constant content of the padding bath, constant wet pick-up, etc.). Also, the fabric has to be carefully and uniformly prepared, and even the moisture content of the fabric should be the same throughout.

12.5 Continuous Dyeing with Reactive Dyes

Two-bath Method

dye/pad/dry/chemical-pad/steam

In this method, the dyes and the base are applied separately. The fabric is padded first with the dye solution containing the proper dyeing assistants, without the base. Included in the padding bath is an antimigrant to help prevent the dye solution moving from wet parts to dry parts on the impregnated fabric, during the following drying step. After careful drying, the fabric is padded a second time with a solution containing the base and a large amount of salt. The presence of an electrolyte will prevent the dye from contaminating the chemical-pad. Then, without drying, the fabric is introduced into a steamer (saturated steam at 215-220°F) for 30 to 60 sec. The fabric is then thoroughly after-scoured by passing it through a range of several wash boxes, and dried or treated further with finishing agents.

The main advantage of this method is that the dye-bath without the base is stable and can be kept overnight without a noticeable change.

One-bath Methods

1. pad/dry/steam

In this method the reactive dyes and the base are applied together from the same padding bath. Two separate solutions are prepared; one contains the dyes, and the other contains the base and other dyeing assistants. The two solutions are mixed according to the amounts needed and fed into the dye pad with the aid of a mixing pump. Without drying, the fabric is introduced into a steamer for 30 to 60 sec. The fabric is then after-scoured and the dyeing process is completed as needed.

2. pad/dry/cure

Here too, the dyes and the base are applied together, and a substantial amount of urea is added to the padding bath. Besides helping dissolve the reactive dyes, the urea will maintain some water on the fabric during the curing stage. This moisture is needed for better fixation yields. After careful drying, the fixation is accomplished by introducing the fabric into the curing oven (e. g. dry heat) at 300°F for 1-2 min. The dyeing process is then completed by after-scouring, etc.

12.6 Hetero-Bifunctional Reactive Dyes

To overcome the high loss of dye during application, hetero-bifunctional reactive dyes have been developed. These dyes contain two different reactive groups. Fig. 12.2 is a general formula for reactive dyes containing a monochlorotriazine and a sulfato-ethyl sulfone groups [2]:

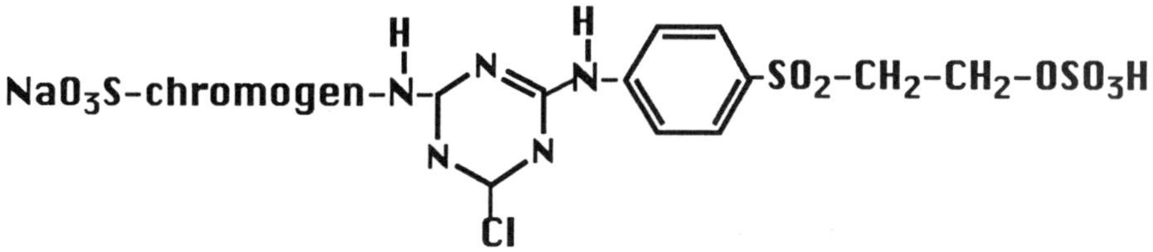

Fig. 12.2 A Bifunctional Reactive Dye

Examples of these dyes are the Sumifix Supra of Sumitomo Chemical [2], and the Cibacron C of Ciba-Geigy. The presence of two reactive groups increases the probability of the dye to bond to the fibers, thus increasing fixation yield.

Each of the covalent bonds formed between a specific reactive group and cellulose has its own drawbacks. Assuming that the dye bonds twice [3], it is also expected that the dye would possess the separate advantages of the two different covalent bonds [2].

References

1. The Dyeing of Fibers with Reactive Dyes, Peter J. Dolby, International Dyeing Symposium, Practical Dyeing Problems, A.A.T.C.C., May 25-26, 1977.
2. Technical Literature of Sumifix Supra Dyes, Sumitomo Chemical, Bifunctional Reactive Dyes, 1990.
3. Some Thoughts on Innovative Impulses in Dyeing Research and Development, Heinrich Zollinger, Textile Chemist and Colorist, Vol. 23, No.12, December 1991,p.19

Chapter 13.

DISPERSE DYES

The cellulose acetate fibers introduced by Celanese Co. in 1920, were the first hydrophobic synthetic fibers. It was soon realized that the existing soluble dyes used for dyeing the hydrophilic natural fibers were not substantive to acetate fibers, and thus could not dye them. The need for a new group of dyes with hydrophobic properties, led to the development of the disperse dyes. These dyes are defined by the Society of Dyers and Colorists as: "substantially water-insoluble dyes having substantivity for one or more hydrophobic fibers, e.g. cellulose acetate, and usually applied from aqueous dispersions".

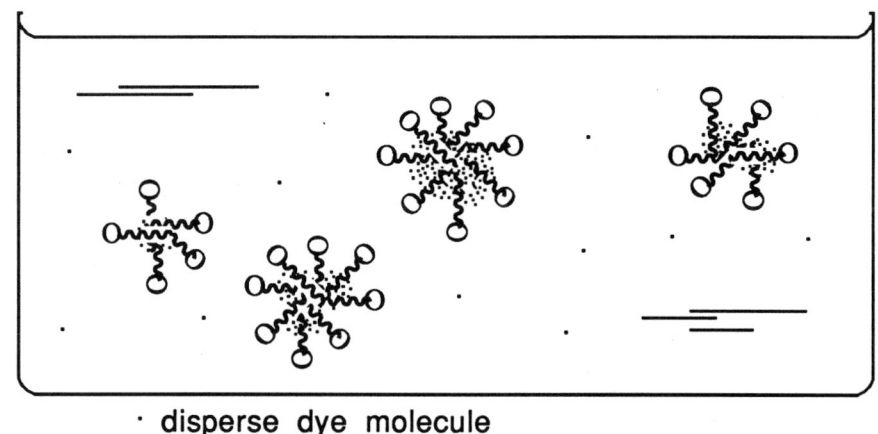

· disperse dye molecule
⌇⌇O dispersing agent

Fig. 13.1 A Dispersion of a Disperse Dye in Water

The disperse dyes are non-ionic aromatic compounds with a relatively low molecular weight, and have an extremely low solubility in water (Fig. 13.1), usually in the range of a few m.g. per liter. Their solubility, however, increases significantly with increasing temperature and in the presence of certain chemicals. Solubility may increase as much as ten-fold at the boil. At 130°C (266°F) (used in pressure dyeing of polyester) and at the usual level of dye concentration, many disperse dyes become almost completely soluble. This minute solubility of the disperse dyes is essential for the dyeing process to occur.

13.1 Commercial Forms of Disperse Dyes

Disperse dyes are marketed in the form of extremely fine particles, about one micron in size, mixed with a dispersing agent(s). These fine particles are obtained by grinding the disperse dye in water together with the dispersing agent for several hours. Anionic compounds, such as sodium- lignine sulfonate or naphthalene sulphonic acid condensates, are usually used as the dispersing agents. Excess water is then removed and the product obtained is processed into its final form as a powder, granular, liquid or paste.

13.2 Dyeing Hydrophobic Fibers with Disperse Dyes

Being hydrophobic, disperse dyes have a good affinity for hydrophobic fibers. Disperse dyes are used primarily in dyeing polyester, cellulose acetate and triacetate, and less frequently in dyeing nylon and acrylic fibers. When penetrating hydrophobic fibers, disperse dyes dissolve in amorphous regions of the fibers. The dyed fiber is referred to as a solid solution, i.e. a solution made of solid components, where the fiber is the solvent and the dye is the solute.

Hydrophobic fibers do not swell readily by merely placing them in water as the hydrophilic fibers do. Therefore, heat must be applied to 'open up' or swell the fiber. Dyeing with disperse dyes is usually performed at temperatures as high as possible without damaging the fibers and/or the dyes. Certain chemicals such as dye-carriers are often used when dyeing polyester or triacetate to assist in 'opening up' the fiber.

In the dye-bath, almost all of the dye is maintained in the form of a dispersion with the aid of a dispersing agent. Only a very small amount of the dye is present as a true solution. This soluble part of the disperse dye, which is in the form of individual molecules, is capable of diffusing into the interior parts of the hydrophobic fiber. As soon as the soluble portion is absorbed, dispersed-solid particles will gradually dissolve to replace the amount of soluble dye molecules lost. In this way a supply of single molecules is continuously available for exhaustion.

The solubility of a disperse dye is affected among other things by its particle size and form of crystallization. The smaller the particle the easier it becomes to dissolve it. Possible disperse dye molecule movements during the dyeing process is shown below:

dispersed dye particles (micelles)
↕
soluble dye (single molecules)
↕
diffusion of dye molecules through the dye-bath
↕
adsorption of dye molecules to surface of fibers
↕
diffusion of dye molecules into the fibers
↕
migration of dye molecules inside the fibers
and back to surface of fibers

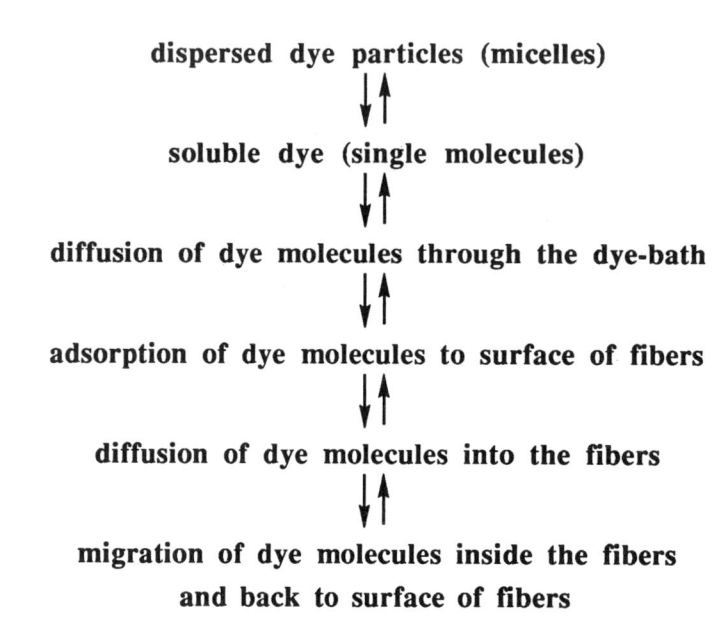

Fig. 13.2 Possible Disperse Dye Molecule Movements During the Dyeing Process

As shown in this diagram re-crystallization of the disperse dye may also take place. This could be the case where the disperse dye is in an unstable crystalline form and during the dyeing process in particular when dyeing at temperatures as high as 135°C (275°F), as in pressure dyeing of polyester, the soluble part of the dye may crystallize into a more stable crystalline form. The resulting crystals may then deposit on the dyed material and form specks. Also the larger the size particle of the disperse dye the higher the tendency of the crystals to grow further.

Factors Influencing the Rate of Dyeing
The rate of dyeing is primarily dependent upon the rate of diffusion of the disperse dyes into amorphous regions of the fibers. In some cases, though, when the number of single dye molecules available is small, the solubility of the dye will become the limiting factor.

Factors affecting the solubility of disperse dyes:
- molecular size and polarity of the disperse dye
- particle size of the dye
- temperature of dyeing
- dyeing assistants (e.g. dye carriers, and certain surfactants)

Factors affecting the rate of diffusion are:
- fiber morphology (T°g and the accessibility of the amorphous regions)
- molecular size of the disperse dye
- temperature of dyeing
- dyeing assistants (e.g. dye-carriers, and certain surfactants)

13.3 Chemical Characteristics of Disperse Dyes

The majority of the disperse dyes are azo benzene derivatives. The second major group, accounting for about 20% of the disperse dyes are the anthraquinone derivatives. Many of the azo disperse dyes, are derivatives of C.I. Disperse Orange 3:

$$O_2N-\langle\bigcirc\rangle-N=N-\langle\bigcirc\rangle-NH_2$$

C.I. Disperse Orange 3
C.I. Solvent Orange 9

Examples of disperse orange 3 derivatives are:

$$O_2N-\langle\bigcirc\rangle-N=N-\langle\bigcirc\rangle-N\begin{matrix}CH_2CH_3\\CH_2CH_2CN\end{matrix}$$

C.I. Disperse Orange 25   (low energy)

$$O_2N-\langle\bigcirc\rangle(Cl)-N=N-\langle\bigcirc\rangle(CH_3)-N\begin{matrix}CH_2CH_3\\CH_2CH_2CN\end{matrix}$$

C.I. Disperse Red 65   (medium energy)

A few disazo disperse dyes are also available, e.g.:

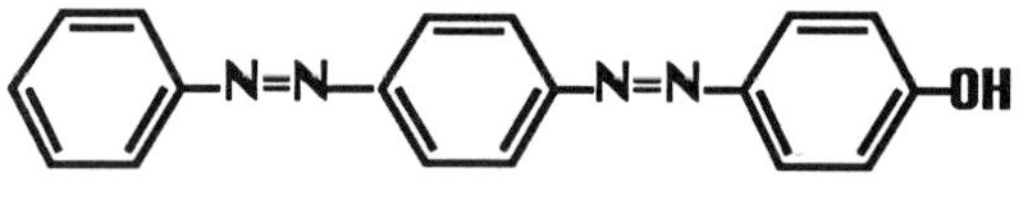

C.I. Disperse Yellow 23   (low energy)

An example of an anthraquinone disperse dye is:

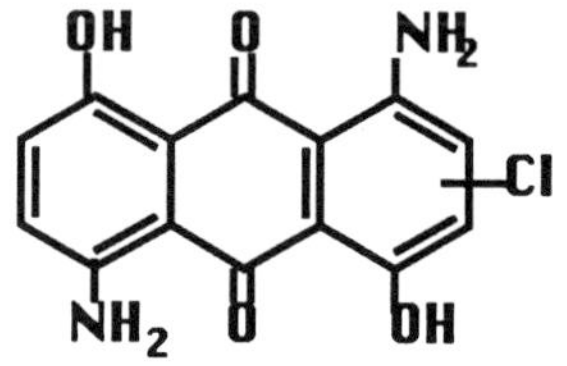

C.I. Disperse Blue 56   (low energy)

Another small subgroup of disperse dyes are nitrophenyl amine derivatives, e.g.:

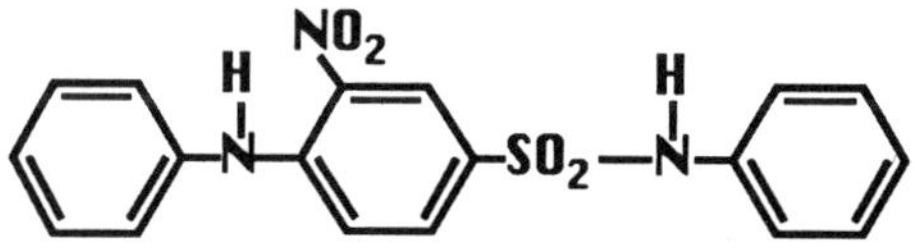

C.I. Disperse Yellow 42   (medium energy)

Most oranges and reds, and some yellows are azo derivatives. Some yellows and oranges are nitrophenyl amine derivatives, which are easy to synthesize and therefore relatively inexpensive. Most violets and blues are anthraquinone derivatives. There are a few blues with azo structures. The few available greens are of special chemical structures.

In general, the anthraquinone disperse dyes, because of their complex synthesis, are more expensive than the azo derivatives. However, level dyeing is easier to achieve with the anthraquinone dyes since they are relatively small in size. Accordingly, the anthraquinone dyes are suitable for light shades (e.g. light blues are almost always produced with anthraquinone disperse dyes). On the other hand, azo disperse dyes, having larger molecular size and being less expensive, are more suitable for producing dark shades where large quantities of dyes are required.

13.4 Classification of Disperse Dyes by their Volatility

Being non-ionic and of relatively small size, disperse dyes tend to sublime at relatively attainable temperatures (160-215°C) (320-420°F). The specific sublimation temperature is dependent upon the size and polarity of the dye molecule. The larger the molecule and/or its polarity, the higher the sublimation temperature. The temperature range in which the dye sublimes defines the dye as a low energy, medium energy, or high energy dye. Low energy disperse dyes have superior leveling properties and are therefore easy to apply. The high energy dyes have the best sublimation fastness but are the most difficult to level. A more detailed classification uses the letters A, B, C, and D, where A is assigned to the lowest energy disperse dyes, and D to the highest energy dyes.

The first disperse dyes, which were called 'Acetate dyes' were developed for dyeing cellulose acetate fibers. Acetate fibers cannot withstand temperatures near the boil. Therefore, the dyes were not exposed to high temperatures during or subsequent to dyeing, and so volatility was of no concern. Typically, the dyes were small molecules which could easily penetrate the fibers at temperatures below the boil.

The limitation of the 'acetate dyes' was soon realized with the introduction of the newer hydrophobic synthetic fibers. After dyeing, these fibers are often exposed to high temperatures such as those used in heat-setting, pleating, or resin finishing of cotton in blends with the hydrophobic fibers. At these high temperatures the dyes tend to sublime leaving dull spots on the fabrics. Furthermore, the evaporated dye condenses on the oven ceiling, and then tends to drop back on top of the fabric that passes below. To overcome this problem, the development of medium and high energy disperse dyes has been pursued.

An example of a medium energy disperse dye is: C.I. Disperse Red 65 (see structure above). Note that this red dye is a derivative of the low energy dye C.I. Disperse Orange 25. The addition of the methyl and chlorine substitutes in C.I. Disperse Red 65 increases the inter-molecular attractions and results in a higher sublimation temperature as compared to that of C.I. Disperse Orange 25.

Examples of high energy disperse dyes are:

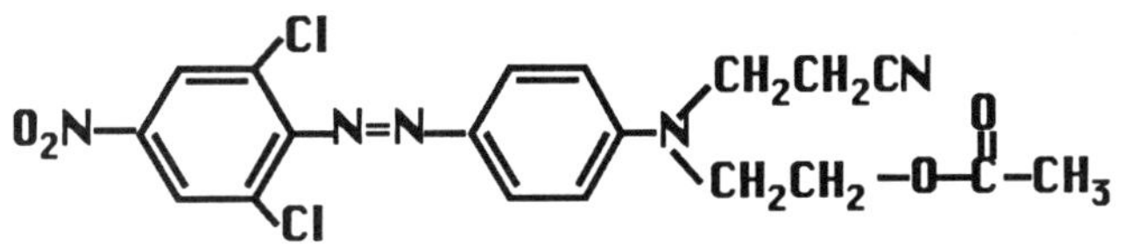

C.I. Disperse Orange 30   (medium-high energy)

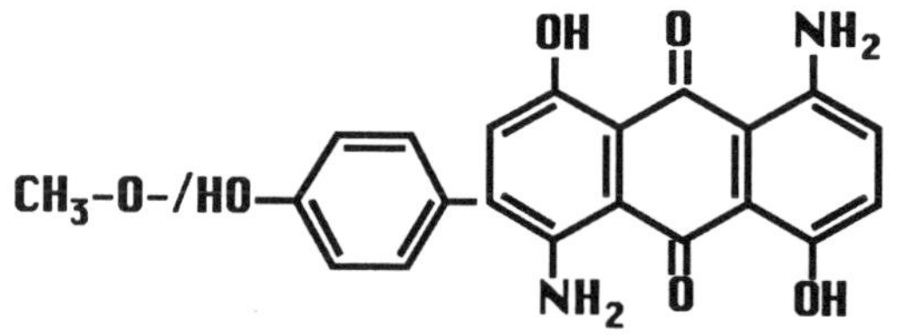

C.I. Disperse Blue 73   (medium-high energy)

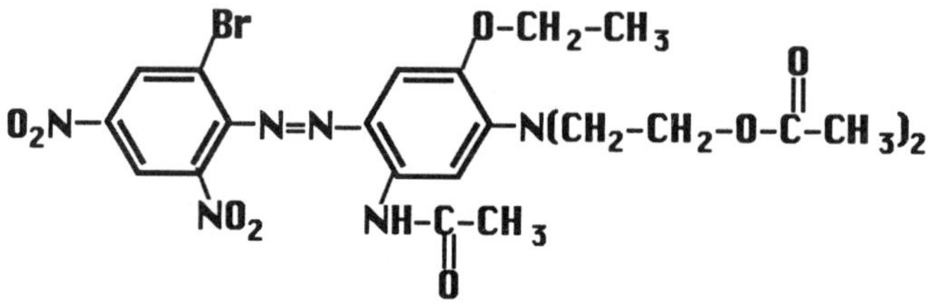

C.I. Disperse Blue 79   (high energy)

Note that the structure of C.I. Disperse Blue 79 is significantly larger and contains an additional polar group as compared to that of C.I. Disperse Blue 56 (See structure above).

13.5 Dyeing Cellulose Acetates with Disperse Dyes

Both acetate and triacetate are chemically modified cellulosic fibers. Triacetate is made by acetylation (esterification) of cotton or wood pulp using acetic anhydride. Triacetate is almost fully acetylated with three acetyl groups per anhydro-glucose unit of the cellulosic chain; hence the name 'triacetate':

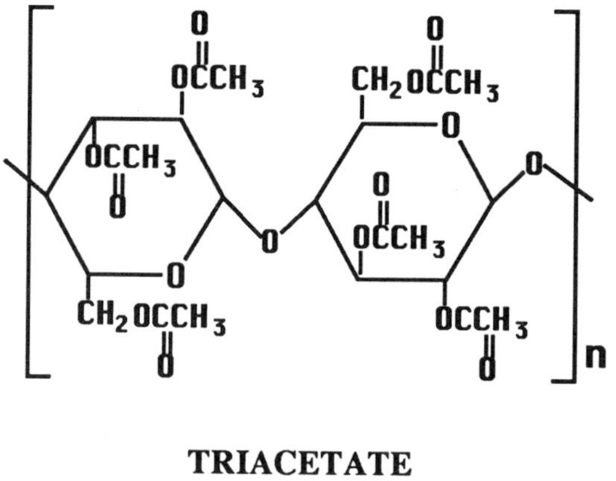

TRIACETATE

Acetate is produced by partial hydrolysis of fully acetylated cellulose triacetate. In acetate there are about 2.3 to 2.4 acetyl groups per anhydro-glucose unit, randomly distributed among the 2,3, and 6 positions.

As a result of the acetylation, the cellulosic fiber becomes hydrophobic in nature. Since acetate has a significant number of 'free' hydroxyl groups left on the fibers, it is less hydrophobic than the fully substituted triacetate.

Acetate fibers are sensitive to saponification (basic hydrolysis of esters), and will hydrolyze back to cellulose when subjected to strong basic solutions at high temperatures. Therefore, in scouring acetate moderate temperatures are used and the pH of the scouring bath is kept below pH 9.5. Triacetate fibers, mainly due to their higher degree of crystallinity, are more resistant to hydrolysis, and thus they can be scoured even at the boil at a pH of up to 9.5.

Acetate and triacetate are thermoplastic fibers. However, heat-setting is performed only on triacetate since dry heat will start causing significant changes to acetate fibers at temperatures as low as 100°C (212°F). Triacetate has a higher melting point and a higher glass transition temperature than acetate. Triacetate fibers have also a more stable and compact structure, and therefore can undergo heat treatments at temperatures as high as 215°C (420°F).

The differences in the physical and chemical characteristics of the two fibers, acetate and triacetate, is believed to be mainly due to the degree of uniformity in the structure of their polymeric molecules. The triacetate molecules having only the hydrophobic acetate groups along their polymeric chains, are more uniform and capable of forming more compact three dimensional structures of higher levels of crystallinity. The acetate polymeric molecules, however, have a mixture of the hydrophobic acetate groups and the hydrophilic hydroxyl groups along their polymeric structure. Therefore, the acetate polymeric molecules with a lower degree of uniformity, forms fibers with lower levels of orientation and crystallinity.

13.6 The Dye-Carriers

Dye-carriers are hydrophobic organic compounds of relatively small molecular size. They enhance swelling of hydrophobic fibers. They are used on fibers that are difficult to penetrate, such as polyester, triacetate, and nomex. While it may take several hours of dyeing at the boil to obtain a light shade on polyester, the inclusion of a carrier will allow the disperse dye to easily diffuse into the fiber, and facilitate the dyeing within an hour.

Examples of dye-carriers are:

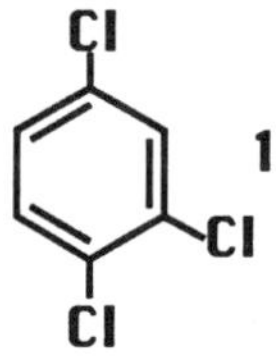

- 1,2,4-trichlorobenzene
- biphenyl 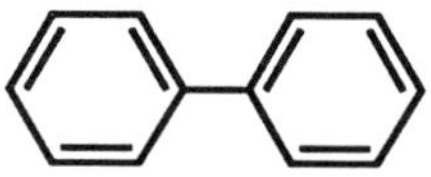
- o-phenyl-phenol 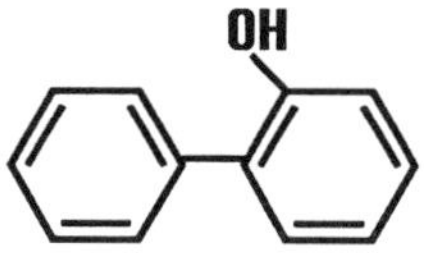
- methyl-naphthalene 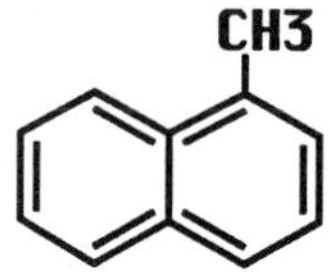

Fig. 13.3   Common Dye-Carriers

The most generally used carriers are the chlorinated benzenes and biphenyl type that are rather inexpensive and do not significantly effect light-fastness if left on the fiber.

Another group of carriers, less effective but also less toxic, are esters of certain organic acids such as:

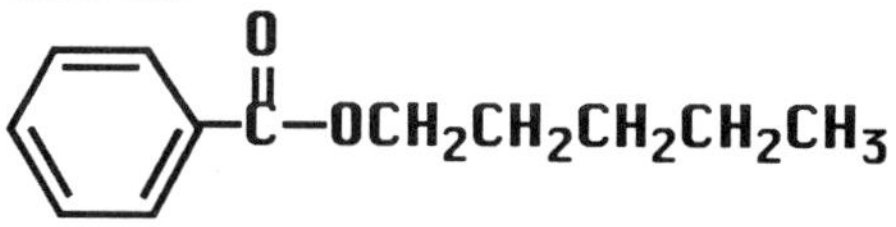

butyl benzoate (C₆H₅–C(=O)–OCH₂CH₂CH₂CH₂CH₃)

Dye-carriers increase the rate of dyeing by affecting both the fiber and the dye-bath. Being more hydrophobic and of a smaller molecular size than that of the disperse dyes, the carriers will penetrate easily into amorphous regions of the fiber, forcing polymeric chain molecules apart. This facilitates the diffusion of large dye molecules into the new spaces formed. In addition, dye carriers increase the hydrophobicity of the dye-bath which enhances the solubility of the disperse dyes, and generates a larger pool of single dye molecules available for exhaustion.

The main drawbacks of using dye-carriers are their toxicity and unpleasant odors. This requires additional efforts, such as thorough scouring and heat treatments, in order to ensure complete removal of the dye carriers. Furthermore, these removal treatments must be performed in compliance with ecological considerations, due to the toxicity and odor of the carriers. Indeed, this drawback is the main reason for the decline in the use of dye-carriers over the past years.

In addition to effectiveness, other factors to be considered, when choosing dye-carriers are: volatility, their emulsifying compound, and their effect on dye-fastness. For example, ortho-phenyl-phenol and methyl naphthalene are highly effective dye-carriers but can cause significant reduction in light-fastness of certain dyes, if they are not completely removed.

For best results dye-carriers are frequently sold as mixtures. Different types of emulsifiers are used to keep the dye-carriers properly emulsified. The choice of an emulsifier should be taken into consideration as sometimes the emulsifier may interfere with dyeing procedures.

Dye-carriers are also used sometimes as leveling agents. Even when accelerated exhaustion is not necessary, small amounts of a dye-carrier may help in opening the fibers more, enhancing the rate of dye-migration.

Using a carrier in concentrations larger than that recommended will result in decreasing the yield of dyeing. Indeed, large amounts of dye-carriers are used for partial stripping of disperse dyes. The dyed material is stripped in a bath containing a large amount of the dye-carrier under similar conditions to those used in the dyeing process.

13.7 Dyeing Acetate Fibers

Disperse dyes are the dyes of choice for acetate fibers, while sometimes, azoic dyes (e.g. diazotized and developed disperse dyes) are used when dark shades are desired.

Acetate fibers are sensitive to high temperatures and start loosing their luster when treated in water at temperatures above 85°C (185°F). Acetate fabrics should be handled carefully and excessive strain or tension should be avoided during their wet processing. In particular, fabrics made of filament yarns should be dyed in the open width (e.g. on a jig or a beam) since the filaments tend to break.

In a typical dyeing procedure the dye-bath is prepared at 120°F, with the pre-dispersed dyes and a dispersing agent if necessary. Acetic acid is added to pH 6-7. The acetate material is placed in the dye-bath and the temperature is raised to 80°C (176°F) within 30 min. Dyeing continues at 80-85°C (176°F-185°F) for one to two hours. The dyeing is followed by rinsing with warm and cold water, and drying.

The disperse dyes used on acetate fibers are of the low energy type. Among them are dyes that have been found to be affected by oxides of nitrogen (e.g. N_2O) present in gas fumes and fumes from industrial areas. Specific examples of these dyes are anthraquinone derivatives containing primary amino groups. The reaction of nitrogen oxides with these dyes results in fading or bleaching of the color. Fabrics dyed with a sensitive disperse dye, in particular one of a blue or violet color, are treated with a certain chemical that acts as an inhibitor to prevent the fading. Gas fading inhibitors are applied together with the dyes or as an after-treatment, according to their proper recommended use.

13.8 Dyeing Triacetate Fibers

Similar to acetate, triacetate is almost always dyed with disperse dyes. Since triacetate has a more compact structure than acetate, higher temperatures are required to open up the fibers sufficiently. The dyeing of triacetate is carried out at temperatures close to the boil, or preferably above the boil at temperatures up to 120ºC (248ºF).

Building medium and deep shades is difficult when dyeing at the boil, so dye-carriers are used to shorten the time of dyeing. The carriers used are of the butyl benzoate type, which are not as effective as those used for polyester (e.g. chlorinated benzene derivatives), but are less toxic and unpleasant to handle as those used for polyester. However, when dyeing triacetate under pressure, even dark shades can be reached within a reasonable time without the use of a dye-carrier.

Triacetate fabrics are heat-set to improve dimensional stability, or to obtain durable pleats. Heat-setting significantly decreases the dyeability of the fabric, but when performed after dyeing, an improvement in wash-fastness is often observed. In addition, when the fabric is heat-set before dyeing, uneven heat-setting can lead to unlevel dyeing. Because of these considerations, it is best to heat-set triacetate fabrics after dyeing.

Chapter 14.

DYEING POLYESTER

The groundwork for the development of polyester fibers was introduced by W. H. Carothers of Du Pont back in 1928. Carothers' interest, however, shifted to the production of nylon which was more promising at that time. The first polyester fiber with proper properties was synthesized by J. R. Whinfield and J. T. Dickson of Calico Printers Ass., but because of World War II and other reasons, it was not until 1948 that polyester filament yarns were first marketed by Imperial Chemical Industries, Ltd. (I.C.I.).

Polyesters are marketed under different trade names. Some of these are: Terylene (I.C.I.), Dacron (Du Pont). Kodel (Eastman), Fortel (Celanese), and Trevira (Hoechst).

Common polyester fibers are made from polyethylene teraphthalate, which is synthesized by condensation polymerization (trans-esterification) of dimethyl-tera-phthalate with ethylene-glycol:

$$CH_3-O-\overset{O}{\overset{\|}{C}}-\underset{\bigcirc}{}-\overset{O}{\overset{\|}{C}}-O-CH_3 + HO-CH_2-CH_2-OH \xrightarrow[H^+]{\Delta}$$

$$H-O-\overset{O}{\overset{\|}{C}}-\underset{\bigcirc}{}-\overset{O}{\overset{\|}{C}}-O-\left[CH_2-CH_2-O-\overset{O}{\overset{\|}{C}}-\underset{\bigcirc}{}-\overset{O}{\overset{\|}{C}}-O\right]_n CH_2-CH_2-OH$$

POLYESTER (polyethylene teraphthalate)

Also available to a small extent are polyester fibers made from other dicarboxylic acids. The long and linear polymeric molecules of this polymer which are more or less planar, can align themselves closely next to each other to form a tightly packed structure with a high degree of crystallinity.

Polyester is hydrophobic, and though it can be dyed with several classes of dyes, regular polyester is dyed only with disperse dyes.

Polyester fibers are strong and have a very good abrasion resistance. They also possess very good dimensional stability and crease resistance, if properly treated. Because of their outstanding properties and their high compatibility in intimate blends with cotton, polyester fibers have become the most widely-used synthetic fibers. Common blends are 50/50 and 65/35 polyester/cotton, which are used extensively in apparel, sheeting, etc.

Basic Dyeable Polyester

To overcome the disadvantages of disperse dyes on polyester, in particular improper wash-fastness due to thermo-migration, the so-called 'basic dyeable polyester' was introduced. This is a modified polyester containing sulphonic acid groups in its polymeric structure serving as dye sites for cationic (basic) dyes. Sulphonic acid groups are incorporate into the polymer by using a small amount of a sulfonated phthalic acid as a co-monomer. Basic dyeable polyester, however, is an expensive fiber and its future use on a large scale is still in doubt.

14.1 Heat-Setting of Polyester

Heat-setting is an essential step in the processing of fabrics containing polyester since unset polyester fabric may shrink by up to 7% when treated in boiling water. Polyester containing yarns are usually heat-set with saturated steam in autoclaves. Heat-setting polyester blends is almost always performed with dry heat on a tenter frame. In the tenter the fabric is exposed to temperatures in the range of 185-215°C (365-419°F) for 30 to 90 sec. Another alternative is to use the perforated drums, but in order to hold the fabric to fixed dimensions it is passed through a tenter before entering the oven. It is possible to heat-set polyester with super-heated steam at lower temperatures (~177°C) (350°F), but this is not widely used, probably because some degradation of the fibers may take place.

In general, heat-setting of polyester decreases its dyeability (dye uptake). The effect of a given heat-setting temperature on the dyeability varies from one type of polyester to the other. Again, in general the dyeability decreases with increasing heat-setting temperature up to ~177°C (350°F). Past this point, however, the dyeability increases with increasing heat-setting temperature and at approximately 205°C (401°F) the dyeability reaches the same level as that of the unset fabric [1].

When needed, polyester fibers can be heat-set again by treating them at or above the previous heat-setting temperature.

14.2 Thermo-Migration

Heat treatments of dyed polyester can cause dye migration to fibers' surface. For example, dyed fibers that initially passed a wash-fastness test, may fail the test after exposure to a temperature of 170°C (338°F) for 30 sec. When exposing the fabric to higher temperatures, such as those applied in heat-setting, the degree of thermo-migration decreases. Migration of disperse dyes may also occur on dyed polyester that was stored for a long time. Residues of dyeing assistants, in particular surfactants may enhance thermo-migration. In addition to wash-fastness other fastness properties, such as fastness to crocking and light, may also be impaired as a result of thermo-migration.

Durable press treatments of polyester/cotton fabrics that follow dyeing can cause thermo-migration. Both, the high temperature applied during curing and the chemicals used in the application may cause dye migration as well as sometimes change of shade. Therefore, dye-fastness tests of polyester/cotton fabrics should be carried out after applying all finishing treatments.

14.3 Dyeing Procedures for Polyester

As previously mentioned, dyeing polyester with disperse dyes at the boil for a reasonable length of time will produce only very pale shades. The compact structure of polyester does not open up to a significant degree when immersed in boiling water, and more vigorous conditions are required for proper dyeing to take place. Good results with disperse dyes are obtained when dyeing by one of the following methods:

-Atmospheric Dyeing with dye-carriers

-High Temperature Dyeing (HT) (Pressure Dyeing)

-Continuous Dyeing by the thermosol process

Atmospheric dyeing is carried out at or near the boil, and dye-carriers must be used to open up the fibers properly. Dyeing under pressure is carried out at temperatures as high as 135°C (275°F). At these temperatures dyes can easily diffuse into the polyester fibers without the use of dye-carriers. In continuous dyeing the fabric is impregnated with the disperse dyes, dried, and then cured at temperatures in the range of 205°C (400°F) for less than a minute. At these temperatures the polyester fiber is open to accept disperse dyes that are already on the fiber's surface.

In exhaust dyeing, when dyeing pastel shades, lower energy disperse dyes should be considered. Level dyeing with these dyes is easier to achieve since they are made of a relatively small molecular size. When dyeing dark or heavy shades, where large quantities of dyes are needed, and level dyeing is easier to attain, relatively inexpensive dyes should be considered. Obviously, for the proper selection of dyes other factors, such as dyeing method, and sequence of fabric processing, must also be taken into consideration.

14.4 Oligomers

During the synthesis of polyester, oligomers made of a few monomers, are also formed. The amount of oligomers in polyester fibers is in the range of 1%. The major product among the oligomers obtained is a cyclic trimer, and therefore the term trimer is often used for these oligomers. When dyeing at high temperatures (e.g. 130°C) (266°F) for a long period of time, oligomers in polyester fibers may be released into the dye-bath. If the oligomers are not kept dispersed in the dye-bath, they will deposit on fibers' surface and cause unlevel dyeing. Oligomers may also form with time heavy deposits on dyeing machines, which are difficult to clean.

During dyeing under pressure, released oligomers remain in the dye-bath in the form a fine dispersion. Upon cooling, the stability of the dispersion decreases, and oligomers start to precipitate. Therefore, to avoid deposits of oligomers, after dyeing the dye-bath should be dropped at a temperature as high as possible.

14.5 Exhaust Dyeing

Processing Sequence of 100% Polyester Fabrics
There are three routes to consider for the processing of a polyester fabric:
      1. SCOUR / HEAT-SET / DYE
      2. HEAT-SET / SCOUR / DYE
      3. SCOUR / DYE / HEAT-SET

1. Scour/Heat-set/Dye
Best results are obtained by following this sequence. Scouring first prevents impurities from being heat-set into the fibers. Heat-setting before dyeing prevents excessive shrinking, and the formation of creases and rope marks in the fabric, during the dyeing process. This method allows the use of dyes that are sensitive to high temperatures (e.g. low energy disperse dyes and/or dyes with poor thermal migration). It also allows the use

of a variety of dyeing equipment (e.g. jet dyeing machine). Dyeing last in the rope form will also help reduce undesirable stiffness introduced during the heat-setting process.

This method, however, is the most <u>costly</u> and <u>time consuming</u> route since it requires a drying step between scouring and dyeing. Furthermore, if the heat-setting equipment is not properly regulated (different temperatures from side to side or from side to center) then the resulting non-uniform heat-setting will lead to unlevel dyeing.

Heat-setting before dyeing is not recommended when dyeing dark shades by atmospheric methods. Usually heat-setting decreases dyeability so that the fibers may not be sufficiently open to accept large amounts of dye.

2. Heat-set/Scour/Dye

This is a more cost effective route since scouring and dyeing can be carried out in the dyeing machine with-out a drying step between them. It has the advantages of heat-setting before dyeing as mentioned above, and the scouring can also remove any yellowing that was introduced during heat-setting. This method, however, is only suitable for fabrics with little or no impurities, such as warp knits of 100% polyester. The scouring that follows may not remove impurities that were set in the fibers during heat-setting.

3. Scour/Dye/Heat-set

This is another cost effective alternative where the wet-processing part is not interrupted by heat-setting. This method can be considered for machines where the fabric is handled in 'open width' during dyeing, provided there is no shrinkage that can not be corrected later in the heat-setting step. Low to medium energy dyes and other heat-sensitive dyes may not be suitable since the dyed fabric will be exposed to high temperatures during heat-setting. When using dye-carriers to promote exhaustion or as leveling agents, heat-setting after dyeing will help in the removal of residual carriers left on the goods.

14.6 Atmospheric Dyeing

Only a small percentage of the polyester fibers in use is currently dyed by this method, since it requires the use of large amounts of the undesirable dye-carriers. Dyeing at temperatures as high as 135°C (275°F) is possible with new dyeing equipment that can withstand high pressures. Under these conditions better diffusion and dye-fastness are obtained without using dye-carriers.

Dye selection

For best application results (ease of diffusion and migration), the disperse dyes used in atmospheric dyeing are of low to medium energy type. In this method the fibers are not open enough to allow larger molecules such as the high energy disperse dyes molecules, to properly diffuse into the fibers. The following is a list of typical disperse dyes used in atmospheric dyeing of polyester:

DISPERSE DYE	**ENERGY LEVEL**
C.I. Disperse:
1. Yellow 23	Low
2. Yellow 42	Medium
3. Yellow 54	Low
4. Yellow 64	Low-Medium
5. Orange 25	Low
6. Red 50	Low
7. Red 60	Low
8. Red 65	Medium
9. Red 73	Medium
10. Red 91	Medium
11. Blue 56	Low
12. Blue 35	Low
13. Blue 73	Med-High (only if necessary for shade matching purposes)
14. Blue 60	Med-High (only if necessary for shade matching purposes)

The following is a typical procedure for atmospheric dyeing:
The dye-bath contains in addition to the disperse dyes and a dispersing agent, 4-8% dye-carrier, antifoam if necessary, and a small amount of a sequestering agent. Many disperse dyes are sensitive to a basic pH at high temperatures, therefore as a common practice the pH of the bath is adjusted to pH 4.5 - 5.5 with acetic acid. The fibers are placed in the dye-bath at 50°C (122°F) and the temperature is gradually raised close to the boil. Dyeing continues at a temperatures near the boil for approximately 1.5 hours., after which the fibers are after-scoured, and dried.

Reduction Clearing

When dyeing medium to heavy shades it is a common practice to include a 'reduction clearing' step in the after-scouring of the fibers. 'Reduction clearing' is very effective in removing disperse dyes from the surface of polyester fibers. In this process, the dyed fibers are placed in a bath containing a reducing agent and a base (e.g. 2g/L hydrosulfite and 1g/L sodium hydroxide). The reducing bath temperature is raised to 80°C (176°F) and kept at this temperature for 15 to 20 minutes.

The effect of reduction clearing depends on the chemical structure of the dye. An azo disperse dye looses its color through the reduction of its azo group to two amino groups (see: stripping of azo dys). An anthraquinone disperse dye becomes water-soluble since the reduction clearing treatment is similar to the solubilizing process of a vat dye. Once in the soluble form the anthraquinone disperse dye will not have any substantivity to the hydrophobic polyester, and will remain in the reducing bath.

When dyeing the polyester component in a polyester/cotton blend, reduction clearing is also very effective in removing disperse dyes that stain the cotton in the blend. Again, the azo dye is destroyed, but one would expect the solubilized anthraquinone disperse dye to be substantive to the cotton like a solubilized vat dye. However, the molecular size of a disperse dye is too small to be substantive to cellulose as the vat dyes are in their solubilized form.

Note that a reduction clearing does not have any effect on the fixed disperse dye, that is, the dye that is inside the fiber. Under the reduction clearing conditions, the polyester remains in its compact structure and the bath content cannot enter the fibers. Also, the reduction clearing reagents (the reducing agent and the base) are ionic and therefore have no affinity to the polyester.

14.7 Pressure Dyeing

Best results with exhaust dyeing are obtained when carried out at temperatures in the range of 130-135°C (266-275°F). Dyeing under these conditions gives:
1) Good overall fastness (wash fastness, crocking, etc.),
2) Particularly good leveling results,
3) The opportunity to eliminate completely the use of dye-carriers with their associated unpleasant properties, and
4) The possibility of using a wider range of disperse dyes, and a better coverage of barre.

Dyeing under pressure at lower temperatures (e.g.115°C) is not recommended. It should only be used when the equipment is not capable of handling the pressure at 135°C (275°F). However, for pastel shades lower temperatures (115°C or even atmospheric dyeing) may be considered.

A small amount of a dye-carrier is some times used for leveling purposes.

Dye Selection

At temperatures in the range of 130-135°C (266-275°F), disperse dyes diffuse easily into polyester fibers without the assistant of dye-carriers. Therefore, medium to high energy disperse dyes are used. These dyes give high yields, have better wash-fastness, and a lower tendency to sublime in subsequent heat treatments.

The following is a list of typical disperse dyes used in pressure dyeing of polyester:

DISPERSE DYE	**ENERGY LEVEL**
C.I. Disperse	
1. Yellow 42	Medium
2. Yellow 54	Low (light to medium shades only)
3. Yellow 64	Medium
4. Orange 29	Medium-High
5. Orange 30	Medium-High
6. Red 60	Low (light to medium shades only)
7. Red 65	Medium
8. Red 73	Medium
9. Blue 60	High
10. Blue 73	Medium-High
11. Blue 79	High
12. Brown 1	Medium-High

Rapid Dyeing Under Pressure

Rapid dyeing is a procedure in which the temperature of the dye-bath is raised under fully controlled conditions. This procedure allows for the time, during which the dye-bath remains at the maximum temperature (migration and fixation period), to be considerably shortened. As a result, a considerable reduction in dyeing time is attained. In addition, the possibility of heat damage, to the dyes and/or the fabric, is reduced since the time of dyeing at maximum temperature is kept to a minimum.

Rapid dyeing is only possible with machines that can operate under the following conditions:

1. Low liquor ratio (approx. 1:7 to 1:10),
2. Uniform bath temperature throughout, and
3. High dye-bath circulation and/or fast movement of the fabric.

Low liquor ratios are important because they allow for rapid raising (or lowering) of the dye-bath temperature. In addition to shortening the dyeing time, low liquor ratios are more economical with respect to energy and chemicals used.

High rates of circulation of approximately 300-700 gallons/minute and/or moving the fabric at a rate of 400-500 yards/minute can be achieved with jet and other types of modern dyeing machines.

Dye manufacturers provide their customers with rapid dyeing procedures that allow level dyeing to occur from the beginning of dyeing so that only a short time of dyeing at maximum temperature is required. In setting the time/temperature program the dyeing characteristics of all the disperse dyes in the dye-bath should be taken into consideration.

The following is a typical example of rapid dyeing of 100% polyester knit that has been heat-set and that it is going to be scoured before dyeing:

STEP 1. Prescour with a non-ionic surfactant (1-2 g/L) and sodium carbonate (1-1.5 g/L). Scour at 65-70°C (149-158°F) for 15-20 minutes, then rinse with a small amount of acid (acetic acid)

Step 1. Remark

Do not use temperatures above 70°C (158°F). This will avoid attaining the cloud-point of the non-ionic detergent, preventing redeposition of oils on the material. It is important that all base be removed.

STEP 2. Set the bath temperature at 50°C (122°F) and add:
a) 0.5-1.0% anionic dispersing agent
b) 0.25% sequestering agent (e.g. EDTA)
c) 0.0-2.0% dye carrier
d) antifoam if necessary (non-silicone type)
e) acetic acid to pH 4.5-5.0

Step 2. Remarks

2a) The presence of a dispersing agent will also aid in the removal of any residual oils or other contaminants. The dispersing agent mentioned above in step 2a) is in addition to the anionic dispersing agent which is already present with the dye as obtained from the dye manufacturer. Accordingly larger concentrations of dyes require less of the additional dispersing agent (e.g. 3% dyeing may contain as much as 1.5% dispersing agent).

2b) Sodium salts of EDTA are the most commonly used sequestering agents. Other sequestering agents that can be used are HEDTA and DTPA.

Sequestering agents are used to form soluble complexes with various metal ions (e.g. calcium, magnesium, and transition metals) that may interfere with the dyeing. In particular copper and iron may cause shade changes with certain dyes.

2c) When dyeing at 135°C (275°F) a carrier is not absolutely necessary, however, it helps in leveling when added. Some dye manufacturers recommend using certain surfactants (instead of a dye-carrier) as leveling agents. Others do not include any leveling agent or carrier in their formulation.

2d) Defoamers are used to prevent foam formation. Various dyeing machines when used under certain conditions develop a foam. This foam may enter the pump hindering dye circulation and/or causing tangling of the fabric. Dyeing equipment that are filled completely with the dye-bath do not have this problem.

2e) A pH below 4.0 or above 5.5 may adversely affect certain disperse dyes. The sensitivity of disperse dyes varies greatly from one disperse dye to another. Using the narrow pH range listed in step 2e), will enable the use of a wide range of disperse dyes in the shading mixture. Also, polyester fibers are more stable at an acidic pH. Prolonged treatments at high temperatures and at a pH above 7 may harm polyester fibers.

<u>STEP 3.</u> Run the bath for 5-10 minutes and check pH.

<u>STEP 4.</u> Add x% pre-dispersed dyes slowly.

<u>Step 4. Remark</u> The choice of dyes requires that the dyes used, in any combination, exhaust in a compatible manner. Use medium to high energy compatible disperse dyes and consult dye manufacturers' literature in choosing the proper combination.

Extreme care must be practiced to disperse the solid forms of the disperse dyes. Slowly sprinkle the dyes into soft water at 50°C (122°F) (avoid much higher temperatures) while constantly stirring (use 2 gallons of soft water for each pound of dye). After adding all of the dye, continue stirring for a minimum of 10 minutes. This dispersion should then be filtered through a fabric or metal filter before entering the dye-bath.

When using liquids or pastes make sure to stir thoroughly in order to re-mix any settling that may have occurred.

<u>STEP 5.</u> Run the bath for 5 minutes and check pH.

<u>STEP 6.</u> Heat as rapidly as possible to 70°C (158°F).

<u>STEP 7.</u> Heat from 70 to 82°C (180°F) at the rate of 2 degrees C. per minute.

<u>Step 7 Remark</u> The initial exhaustion temperature of the dyes should determine the rate of heating. Those dyes which will exhaust at a higher temperature (high energy type) can be heated at a faster rate at the above temperatures.

<u>STEP 8.</u> Heat from 82°C (180°F) to 130°C (266°F) at the rate of 1.5°C per minute.

STEP 9. Dye at 130°C (266°F) for 20 minutes. Dyeing at 135°C (266°F) will require only 15 minutes.

Step 9 Remark The time of dyeing also depends on depth of shade. For ight shades 10 minutes at maximum temperature should be sufficient. Check the sensitivity of the chosen dyes before raising the temperature above 130°C (266°F).

STEP 10. Cool to 82°C (180°F) and sample. If necessary make 'adds' and run at 130°C (266°F) for an additional 20 minutes.

Step 10 Remark If additions are necessary they should be provided in small amounts and constantly monitored again for dye content, dyeing assistants, pH, etc.

Note, that if you are repeating a dyeing procedure, the need for 'adds' may indicate one or more of the following problems:

1) A change in the nature of fibers or improper preparation.
2) An error in the weighing of the dyes and/or dyeing assistants.
3) Contamination in the water used.
4) A defect in the dyeing equipment (leaking pipes or a failure in the temperature and circulation controls).

STEP 11. Cool to 80°C (176°F) and rinse until clear.

Step 11 Remark Make certain that the temperature to which the bath is lowered is sufficiently high to prevent the deposits of oligomers (trimer) and/or other ingredients onto the goods.

STEP 12. After-scour (not needed for light to medium shades)
-Reduction Clearing-
a) set bath at 50°C (122°F)
b) add 2% sodium carbonate (or 0.5% sodium hydroxide)
c) raise the temperature to 77°C (170°F)
d) add 2-3% sodium hydrosulfite
e) run at 77°C-82°C (170-180°F) for 15-20 minutes

Step 12 Remark An alternative to sodium hydrosulfite is thiourea-dioxide. Although it is more expensive than hydrosulfite, it is more stable and requires only 1/4 or less of the concentration of hydrosulfite.

STEP 13. Rinse with hot water.

STEP 14. Neutralize with acetic acid at 50°C (122°F) for 5-10 minutes

14.8 Continuous Dyeing by the Thermosol Process

This process was first introduced by Du Pont in 1949 (R. H. Thomas, 64th National Meeting, A.I.C.E. New Orleans, LA, 1969), and is based on the idea of dissolving disperse dyes in hydrophobic fibers, mainly polyester, by means of dry heat. The application is carried out continuously and consists of padding the polyester fabric with disperse dyes, drying, and then placing the fabric in a curing oven at high temperatures (200-210°C) (392-410°F) for less than one minute, during which time the dyes diffuse and dissolve in the polyester. At temperatures in the range of 200°C (392°F) the polyester fibers are sufficiently open to allow a rapid diffusion of the disperse dyes. The thermosol process is called so because the solid solution of the dye in the fiber is achieved with the aid of heat.

The padding bath contains in addition to the dyes, acetic acid (pH 4.5-5.5) and a wetting agent (1-3g/L). A dispersing agent is added only if necessary, since the dyes themselves contain dispersing agents as fillers. The dispersing agent and the wetting agent are weakly attracted to the disperse dyes and tend to hold them back from moving into the polyester during the thermosol stage. Therefore larger amounts of these agents may reduce fixation yield. For the same reason, dyes in liquid form are preferred since their dispersing-agent content is significantly lower than that of other forms of disperse dyes (e.g. powder or granular forms).

During the thermosol stage disperse dyes move into the polyester fiber through contact migration and/or sublimation. That sublimation of disperse dyes actually takes place can be shown in the thermosol dyeing of polyester/cotton blends. Since polyester is hydrophobic, during padding most of the dye dispersion is being picked up by the cotton component in the blend. However, after thermosoling most of the dye is found to be on the polyester. This clearly indicates that transfer of dye from the cotton in the blend to the polyester took place through sublimation. Indeed the disperse dyes used for polyester/cotton fabrics are of the high energy type that tend to sublime at the thermosol temperature.

The transfer printing process is also based on the volatility of the disperse dyes. In this process, a fine design is printed first with disperse dyes on the transfer paper. The print is then transferred by pressing the printed paper and the polyester fabric together under heat. At the pressing temperature (~210°C) (410°F) the volatile disperse dyes sublime and the colored design is absorbed by the polyester.

In addition to great savings in time when performed on a large scale, continuous dyeing yields dyed goods with very good overall fastness. In particular, very good barre coverage can be achieved. However, fibers may lose most of their crimp and the hand of the fabric may be inferior as compared to that of jet dyed fabrics.

14.9 Stripping Disperse Dyes From Polyester

As was mentioned in the previous chapter, partial stripping can be obtained by treating the dyed fibers in a bath containing a large amount of dye-carrier (10% o.w.f. and more).

Oxidizing or reducing agents can also be used for stripping disperse dyes. However, these treatments may cause damage to fibers and therefore their application conditions must be carefully planned.

Sodium chlorite (0.2% o.w.f.) can be used as an oxidizing agent. The application is carried out at the boil in presence of dye-carriers. Hydrosulfite in presence of sodium hydroxide can be used as a reducing system. The application is carried out at the boil in presence of a carrier, or at temperatures above the boil. This system is very powerful, but because of the presence of alkali it can damage the polyester.

References

1. Marvin D.N., J.S.D.C., 70, 1954, 16.
2. Dyeing and Finishing of Polyester Fibers, Manual, BASF
3. Polyester Finishing, SANDOZ LTD
4. Finishing of Polyester Fibres, BAYER FARBEN REVUE, 1983
5. Dyeing and Finishing of Polyester Fiber, OKSAKA SENKEN LTD., JTN (Japan Textile News, Osaka 541, JAPAN, 1978.

Chapter 15.

DYEING NYLON FIBERS

The polyamide Nylon 6,6 was the first true man-made fiber, to be offered to the public on a commercial scale. It is a fully synthetic fiber made from simple monomers. The fundamental studies on fiber-forming polyamides were initiated by Wallace H. Carothers at the laboratories of E.I. Du Pont de Nemours and Co. in the year 1928. On October 27, 1938, the discovery of nylon 6,6, was announced to the public by Du Pont. The name nylon had been coined as a generic name for synthetic polyamide fibers.

15.1 Chemical and Physical Properties of Nylon Fibers

The most common nylons are nylon 6,6 and nylon 6. They are made by condensation polymerization of bifunctional monomers containing amino and/or carboxylic groups. The number in the name indicates the number of carbons in the monomer(s) from which the fiber is made.

Nylon 6,6 is synthesized from hexamethylene-diamine: $H_2N-(CH_2)_6-NH_2$, and adipic acid: $HOOC-(CH_2)_4-COOH$, and has the following structure:

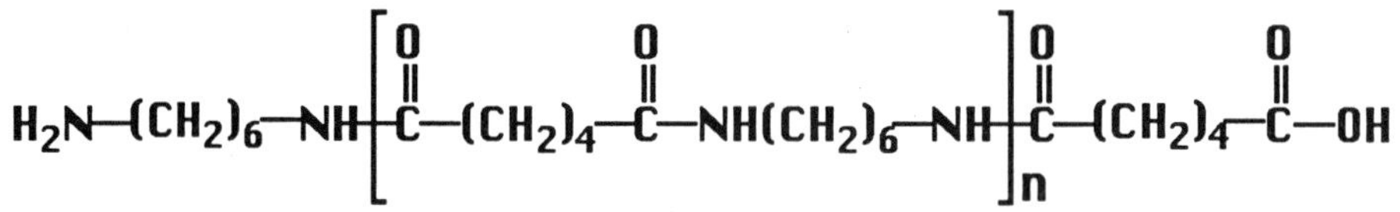

$$H_2N-(CH_2)_6-NH \left[\overset{O}{\underset{\|}{C}}-(CH_2)_4-\overset{O}{\underset{\|}{C}}-NH(CH_2)_6-NH \right]_n \overset{O}{\underset{\|}{C}}-(CH_2)_4-\overset{O}{\underset{\|}{C}}-OH$$

NYLON 6,6

Nylon 6 is made from the monomer caprolactam. In the course of the synthesis, the caprolactam ring opens to 6-amino-hexanoic acid: $H_2N-(CH_2)_5COOH$. Then, this amino acid undergoes self condensation to yield the polymer nylon 6:

$$H_2N-(CH_2)_5-\overset{\overset{O}{\|}}{C}-\left[NH(CH_2)_5-\overset{\overset{O}{\|}}{C}\right]_n NH(CH_2)_5-\overset{\overset{O}{\|}}{C}-OH$$

NYLON 6

Note that in each polymeric chain there is an amino group at one end of the chain and a carboxylic group at the other end. These functional groups may serve as dye sites for certain dyes.

The aromatic polyamides such as Nomex and Kevlar are also nylons by definition. However, their dyeing properties vary greatly from those of the common nylons. Both fibers have highly compact structures with a high degree of crystallinity. Nomex, for example can be dyed with disperse dyes only under pressure , at 130ºC, and in the presence of dye carriers. Kevlar cannot be dyed properly even under these conditions. These fibers will not be discussed further here, and the term nylon will be used to refer to the common aliphatic nylons such as nylon 6, and nylon 6,6.

Nylon fibers have high tensile strength and abrasion resistance, and very good elastic properties. They have good resistance to weather, bases, and bleaching agents. Nylon fibers are less stable in acidic solutions and may undergo changes even when exposed to mild acidity such as acidic air pollution. Prolonged treatments in basic solutions, however, may also cause hydrolysis of the polymer.

Nylon fibers are used extensively in carpets, upholstery, apparel, and in out-door fabrics, such as tents and automotive upholstery. Nylon is also used in intimate blends with other fibers to provide strength. A very attractive combination is the nylon/wool blend used in manufacturing high quality upholstery fabrics.

Even though nylon is basically hydrophobic, hot water will penetrate the fibers and open up the amorphous regions. Even at room temperature, the moisture content of nylon is about 4%, ten times as high as that of polyester. The attraction between water and nylon is in part due to the presence of some ionic sites on the fiber. When an amino group from one polymeric chain is close enough to a carboxylic group of an adjacent chain, the two will react to form an ionic bond (inner salt). The number of these ionic bonds is very small, yet enough to make it more attractive for water to enter the fibers.

The two fiber types, nylon 6,6 (m.p. 250ºC) (480ºF), and nylon 6 (m.p. 215ºC) (420ºF), differ in their dyeability. Nylon 6 has a higher degree of amorphous regions, and therefore diffusion and level dyeing are easier to attain. On the other hand better wash-fastness is achieved on nylon 6,6.

15.2 Heat-Setting of Nylon

Nylon can be heat-set with dry heat, steam, or even with boiling water. Whether performed before or after dyeing, heat-setting must be applied uniformly. Once nylon is fully set, it can not be reset by post-setting even with higher temperatures than those used in the previous setting. Thus, for example, unset nylon fabrics must not be dyed at the boil in the rope form (e.g. on a beck). Creases which form at this stage will not be removed by subsequent heat-setting at higher temperatures. To avoid crack marks and shrinking of unset nylon at the boil, nylon fabrics are usually heat-set before dyeing. Therefore the preferred sequence for processing nylon fabrics is: scour / heat-set / dye. In order to reduce expenses, sometimes the fabric is heat-set prior to the wet processing steps.

Heat-setting with dry heat increases crystallinity of the fibers and decreases both the dyeability and rate of dye uptake. However, heat-setting with steam under pressure increases the rate of dye uptake, but does not affect the dyeability of the fibers.

When heat-setting with hot air continuously on a tenter frame, nylon 6,6 is treated at temperatures in the range of 200°C-215°C (392°F-a420°F) for 15-30 sec., while nylon 6, because of its lower melting point, is heat-set at 177°C-193°C (350°F-380°F) for 15-30 sec. At these high temperatures, oxidation of the nylon may occur, resulting in yellowing and stiffening of the fibers. To avoid oxidation, the air in the curing oven is replaced with water by injecting super heated steam.

Heat-setting with saturated steam is performed batch-wise, usually on knits. A roll of the fabric is placed in an autoclave and saturated steam is injected into the chamber. The duration of the treatment takes between 10 to 20 min., during which time the temperature in the autoclave may reach 132°C (270°F).

15.3 Dye Selection

Nylon fibers are unique in that, in principle, they can be dyed with almost all types of dyes. This is so because of the presence of amino and carboxylic groups, the hydrophobic nature of the fiber, and the ability of water to penetrate the fiber. Accordingly, anionic dyes such as acid and direct dyes form ionic bonds with amino groups on the fiber, and cationic dyes form ionic bonds with carboxylic groups on the fiber. Reactive dyes form covalent bonds with the amino groups. Disperse dyes dye nylon in the same way as they do other hydrophobic fibers. Finally, nylon fibers can also be dyed with azoic, vat, and sulfur dyes.

One of the main problems associated with dyeing nylon is unlevel dyeing which results from irregularities in fibers, and is referred to as 'barré'. This type of unlevelness shows up as lighter or darker streaks or stripes on woven or knitted fabrics extending crosswise. Barré is much more evident when using dyes that depend on the amino or the carboxylic groups as dye sites.

Barré results from chemical and physical variations in nylon fibers. Variations in fibers can be the result of using different conditions during the synthesis of the fibers, spinning, drawing, and texturing. Varying the conditions of these heat treatments can change the number of amino and/or carboxylic groups and/or their accessibility. During a heat treatment, a certain number of amino groups will oxidize. Heat treatments also affect the degree of crystallinity, thus changing the dyeability of the fibers. In addition, uneven tension introduced during the spinning process of making yarns, knitting or weaving, may also lead to unlevel dyeing.

The dyes most widely used on nylon fibers are the anionic dyes (acid, metallized, and some direct dyes). Disperse dyes are used to a lesser extent, mainly for producing light shades. Reactive dyes are used in special cases where brilliant colors with very good wash-fastness are desired.

In general, dyes with low substantivity and good leveling properties will give better coverage of fiber irregularities. However, wash-fastness may not be adequate, and dyes of high substantivity with more complex structures are needed to meet wash-fastness requirements.

15.4 Dyeing Nylon with Acid and Metallized Dyes

The applications of various acid dyes to nylon fibers are for the most part similar to their applications to wool. Since acid dyes are more substantive to nylon, a smaller amount of acid is required for exhaustion as compared to the amount used in dyeing wool. Thus, acid leveling dyes are applied at a pH of about 5, whereas neutral dyeing acid dyes, 2:1 metallized dyes, or selected direct dyes are applied at a neutral pH or slightly above pH 7. Because of the higher substantivity, the wash-fastness of acid dyes on nylon is better then that obtained for wool. Level dyeing increases with increasing temperature, and dyeing is usually carried out near the boil. The rate of exhaustion is relatively high and therefore the dyeing temperature is raised gradually under controlled conditions.

The number of amino groups in nylon fibers is much smaller (20 times less) than that of wool. When dyeing nylon with acid dyes, a saturation point is sometimes reached before the desired shade is attained. Therefore, in this case it can be difficult to build heavy shades. Monosulfonated dyes are more suitable for dyeing heavy shades as compared to the disulfonate dyes which occupy more amino groups in the nylon fibers.

Acid Leveling Dyes

Selected acid leveling dyes can give good coverage of fiber irregularities, and are relatively easy to level. Both anionic and cationic leveling agents are used to slow down dye exhaustion. The mode of action of these leveling agents is explained in Chapter 7. Dyeing is carried out at temperatures close to the boil for 1-2 hours. However, level dyeing will significantly improve when dyeing under pressure at 116°C (240°F) for a shorter period of time.

In general, acid leveling dyes are easy to level, and can cover irregularities in yarns and fabrics. A wide range of colors with many bright shades is available. Selected acid leveling dyes can provide good wash-fastness and light-fastness with bright colors.

One of the main disadvantages in using acid dyes on nylon is that many of the acid dyes are incompatible when used together in the same bath. Since there is a limited number of amino groups available, the nylon fibers may not have 'enough room' for all the dyes in the mixture, and commonly there will be a preference for a more substantive dye. This phenomena is referred to as the 'blocking effect', where one acid dye is blocking another from attaching to the fiber. Therefore, when dyeing with a trichromatic combination, it is important, to select dyes that exhaust on tone.

2:1 Metallized Dyes

When applied to nylon, 2:1 metallized dyes as well as neutral dyeing acid dyes have a very low tendency to migrate. Therefore, when using these dyes, controlled exhaustion becomes a must. During application, the dye-bath temperature is increased at a slow rate (e.g. 1°C per min.), and anionic leveling agents are used to slow down the rate of dye up-take. Cationic leveling agents should be avoided since they can cause these dyes to precipitate.

2:1 metallized dyes have very good wash and light-fastness, but their shades are not nearly as bright as those of the acid leveling dyes. Therefore, they are used extensively on nylon carpets where bright colors are usually not desired. The 2:1 metallized dyes are also suitable for building heavy shades.

15.5 Dyeing Nylon with Disperse Dyes

Disperse dyes have good affinity to nylon fibers and form with them solid solutions, as they do with other hydrophobic fibers. Since nylon fibers open up sufficiently in water at temperatures close to the boil, disperse dyes are applied to them without the use of dye-carriers. Unlike the acid dyes, the disperse dyes are not dependent on a limited number of dye sites. Therefore, they are usually compatible when used in combinations. For the same reason, they cover fiber irregularities (barre') very well.

During application to nylon, disperse dyes migrate well and are therefore relatively easy to apply. In a typical dyeing procedure, the bath is prepared with the disperse dyes and a dispersing agent, and the pH is adjusted to 5-6 with acetic acid. The nylon material is inserted and the dye-bath temperature is gradually raised to 88°C-93°C (190°F-200°F). Dyeing continues at maximum temperature for about one hour, after which the material is rinsed and dried in the regular manner.

The main problem with disperse dyes is their limited wash-fastness, and hence they are almost always used for light shades. Sometimes poor fastness to crocking also becomes a problem when dyeing heavier shades. Light-fastness varies from fair to very good. Disperse dyes suitable for dyeing nylon are available in a wide range of shades including bright colors.

The disperse dyes that give very good coverage of barre' are of the low energy type, and therefore have poor sublimation fastness. The high energy type dyes, which have good sublimation fastness, do not cover barre' well.

The following are disperse dyes that have very good leveling properties when applied to nylon:

C.I. Disperse Yellow 54	C.I. Disperse Yellow 64
C.I. Disperse Red 1	C.I. Disperse Red 11
C.I. Disperse Blue 3	C.I. Disperse Blue 7

All these are low energy disperse dyes.

15.6 Modified Nylon Fibers

In order to produce multicolored effects on nylon yarns or fabrics, modified nylon fibers with different dyeing characteristics have been developed. Intimate blends of these fibers are used to produce heather effects or yarns with contrasting colors. Sometimes, these type of yarns are used in making multicolored nylon carpets.

Basic Dyeable Nylon

These fibers contain a large number of acidic groups such as sulphonic or carboxylic acid groups. Accordingly, these fibers have an increased substantivity to basic dyes and at the same time a decreased substantivity to anionic dyes. Basic dyeable nylon fibers can be used in blends with acid-dyeable nylon to achieve multicolored effects (cross-dyeing) on 100% nylon . By using selected anionic and cationic dyes, both dye types can be applied to these blends from a single dye-bath.

Deep and Light-Dyeable Nylons

These nylon fibers are substantive to anionic dyes to various extents. The deep-dyeable nylon has a much higher number of amino groups (or other basic groups) than the regular nylon, whereas the light-dyeable nylon contains only a small number of basic groups compared to regular nylon. Accordingly, when dyeing a mixture of these fibers, all fibers will have the same hue but they will differ in their depth of shade.

References

1. Chemical Processing of Synthetic Fibers and Blends, K.V. Date, and A.A. Vaidya, John Wiley & Sons, 1984, Chapter 9, Dyeing of Nylon
2. The Dyeing of Synthetic-Polymer and Acetate Fibers, Edited by D.M. Nunn, The Dyers Company Publications Trust,West Yorkshire, England, 1979, Chapter 4. Dyeing of Nylon and Polyurethane Fibers, by P. Ginns and K. Silkstone

Chapter 16.

DYEING ACRYLIC FIBERS

16.1 Chemical and Physical Properties of Acrylic Fibers

The first acrylic fibers offered by Du Pont in 1948 were homo polymers of acrylo-nitrile ($CH_2=CH-CN$):

$$\sim\!\!\sim\!-CH_2-\underset{\underset{CN}{|}}{CH}-CH_2-\underset{\underset{CN}{|}}{CH}-CH_2-\underset{\underset{CN}{|}}{CH}-CH_2-\underset{\underset{CN}{|}}{CH}-CH_2-\underset{\underset{CN}{|}}{CH}-CH_2-\underset{\underset{CN}{|}}{CH}-CH_2-\underset{\underset{CN}{|}}{CH}-\!\!\sim\!\!\sim$$

Polyacrylo-nitrile

Since polyacrylo-nitrile chains are long and very thin, without bulky side-chains, they can form structures with high degree of crystallinity. Fibers made of these polymer have highly compact structures, with glass transition temperatures above the boil. Therefore, it is very difficult to dye them under atmospheric conditions. By incorporating other co-monomers into the polymer, the regularity of the polymeric chain is interrupted and the degree of orientation and crystallinity decreases. Fibers made of these co-polymers have more open structures with glass-transition temperatures below the boil.

Regular acrylic fibers are co-polymers made of at least 85% (by weight) of acrylo-nitrile. They are synthesized by addition co-polymerization with other monomers such as methyl-methacrylate, methacrylic acid, or sulphonated styrene.

When a fiber contains less than 85% but at least 35% of acrylo-nitrile, it is referred to as a 'modacrylic' fiber. Of special interest are the modacrylics made of acrylo-nitrile and vinyl chloride or vinylidene chloride ($CH_2=CCl_2$). The high chlorine content of these modacrylics impart them flame retardancy properties.

Acrylic fibers are marketed under different names, among them are: Orlon (Du Pont), Acrilan (Monsanto), Creslan (American Cyanamid), and Zefran (Dow Badische).

Acrylic fibers have overall good physical properties, and very good resistance to weathering, and chemicals. Textured acrylic fibers resemble wool in their bulky and soft hand. Since they are relatively inexpensive and can be dyed with brilliant colors, the acrylic fibers are used extensively as a substitute for wool.

The light-sensitive cationic (basic) dyes on acrylic fibers show unexpected good stability to sunlight, therefore, regular acrylic fibers are usually dyed with them. These basic dyeable acrylics contain anionic groups along their polymeric chains which attract the colored cations of basic dyes.

In the synthesis of basic dyeable acrylics, catalysts such as potassium persulfate are used which produce the polymeric chains with sulfonic groups at their ends:

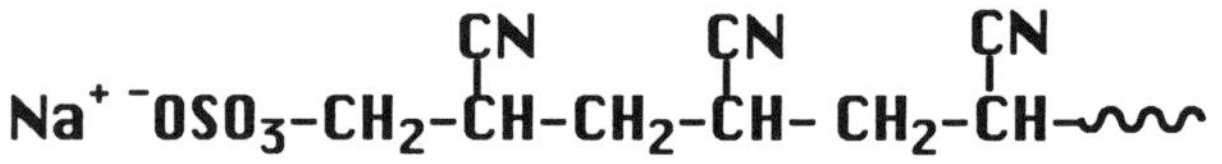

$$Na^+ \ ^-OSO_3-CH_2-\underset{\underset{CN}{|}}{CH}-CH_2-\underset{\underset{CN}{|}}{CH}-CH_2-\underset{\underset{CN}{|}}{CH}\sim\sim$$

Additional anionic groups are added by using co-monomers containing anionic groups, such as methacrylic acid or sulfonated styrene ($CH_2=CH-C_6H_4-SO_3^-Na^+$):

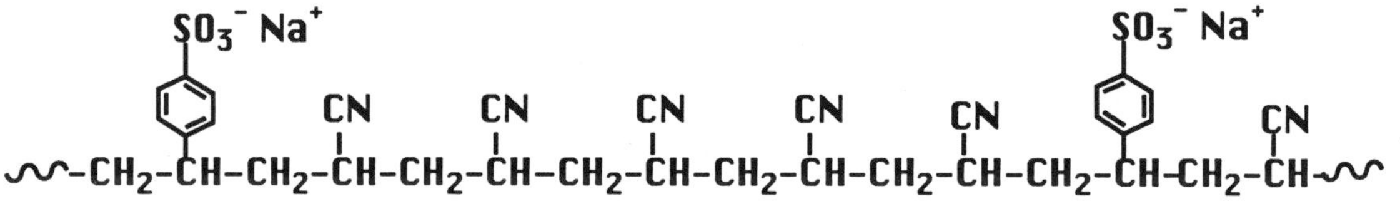

Acrylonitrile-sulfonated-styrene co-polymer

Acrylic fibers have typical glass transition temperatures (T^og), in the range of 160-190°F. Below the T^og, the fibers are not sufficiently open, rate of dyeing is very slow, and only a small amount of dye penetrates them. Above the T^og, a sudden opening takes place, and the dye diffuses into the fibers very rapidly. When dyeing with cationic dyes above the T^og of the acrylic fiber, an increase of several degrees in temperature can double the rate of dyeing.

Available also to a very small extent are the acid-dyeable acrylics which are made to be dyed with anionic dyes. These fibers are manufactured mainly for cross-dyeing purposes. Blends of acid and basic-dyeable acrylic fibers can be dyed with contrasting colors, from a single dye-bath containing acid and cationic dyes.

16.2 Dyeing Acrylic Fibers with Disperse Dyes

The mechanism by which disperse dyes dye acrylics is similar to that of other hydrophobic fibers, i.e., the formation of a solid solution. However, the solubility of disperse dyes in acrylic fibers is limited, and they are used mainly for producing light shades.

When applied to acrylic fibers, disperse dyes show very good diffusion and migration properties. Also, acrylic fibers dyed with disperse dyes in light shades have good wash-fastness. Therefore, although cationic dyes can be used to obtain light shades on acrylics, disperse dyes are preferred since they are much easier to apply.

16.3 Dyeing Acrylic Fibers with Cationic Dyes

Acrylic fibers are almost always dyed with cationic (basic) dyes that yield brilliant colors with excellent wash-fastness. The first basic dyes were applied to natural fibers and showed high sensitivity to light. When it was found that a significant improvement in light-fastness is obtained on acrylics, other basic dyes with even better light-fastness have been developed.

While it is easy to get the dyes into acrylic fibers at the proper temperature, the most difficult task in dyeing with cationic dyes is to achieve level dyeing. This is because of the fast strike that takes place when the dyeing temperature reaches the glass transition temperature of the fibers, and the very strong bonds that are formed between cationic dyes and acrylic fibers. The presence of strong ionic bonds (Fig. 16.1) makes it difficult for migration to take place. It is therefore necessary to exhaust the cationic dyes under controlled conditions, but this is not sufficient by itself. Below the glass transition temperature, the rate of dyeing is extremely low and past this critical temperature, during which time a sudden opening of the fibers takes place, the dyes diffuse very rapidly. Since it is very difficult to correct this initial strike through migration, the rate of exhaustion is also controlled by the use of retarders. The selection of a proper retarder should be made carefully but a full remedy to the problem has yet to be found.

The cationic dyes bond to acrylic fibers by ionic bonds similar to the bonds obtained between anionic dyes and polyamide. The colored cations of the dye combine with the negatively charged sites on the fibers through an ion-exchange reaction to yield the colored fiber:

$$\text{DYE-}\overset{+}{N}(CH_3)\ Cl^- + \text{FIBER-}SO_3^-\ Na^+ \longrightarrow \text{DYE-}(CH_3)\overset{+}{N}\ ^-{}_3OS\text{-FIBER} + NaCl$$

The DYE-FIBER combination thus obtained is shown in Fig. 16.1.

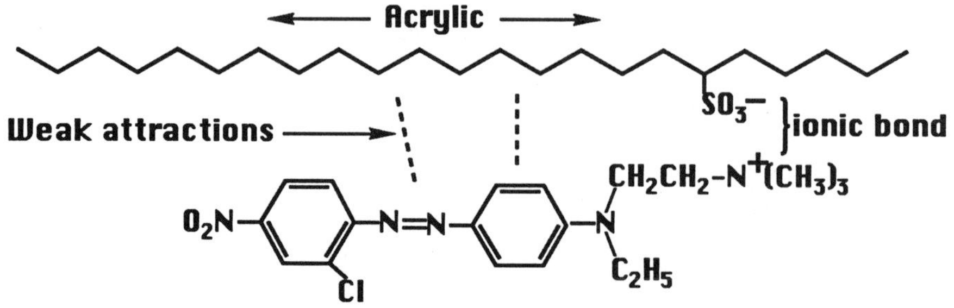

Fig. 16.1 Chemical Bonds Between a Cationic Dye and an Acrylic Fiber

The following are examples of cationic dyes currently in use:

C.I. Basic Blue 3   K = 3.5

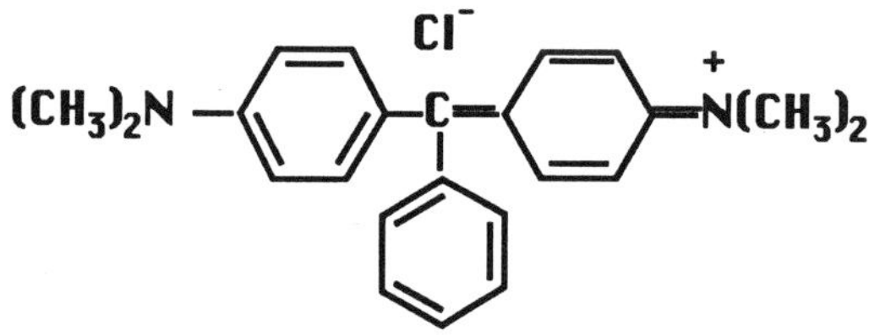

C.I. Basic Green 4   K = 3.5

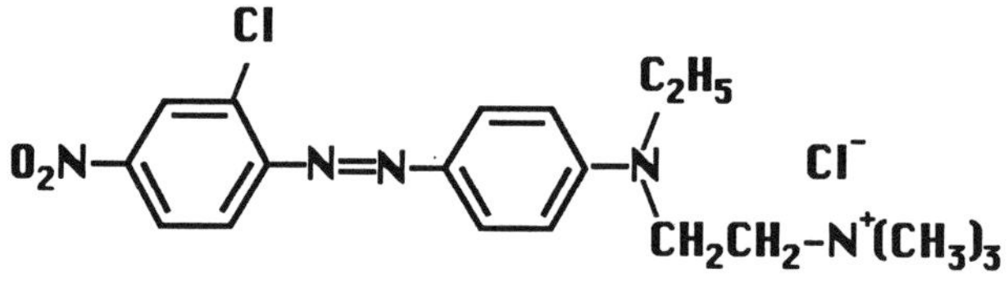

C.I. Basic Red 18   K = 2.5

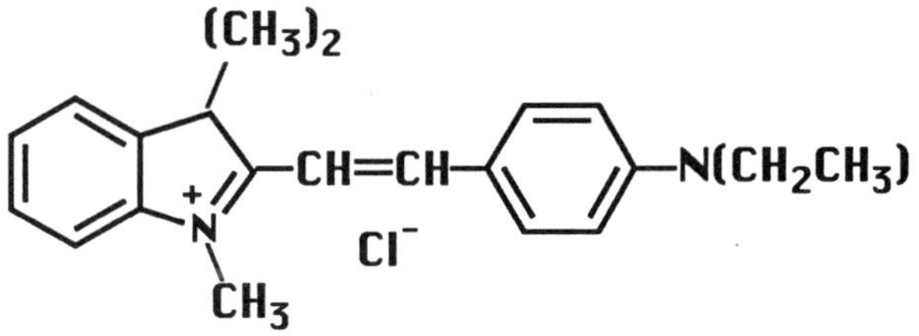

C.I. Basic Violet 16   K = 1.5

The Application of Cationic Dyes

The following is a typical procedure for dyeing a medium shade on a 100% acrylic knit with cationic dyes:

The dye-bath is first prepared with the dyeing assistants at 120°F. Acidic acid is added to obtain a pH of 4.5-5, and sodium sulfate (5-10% o.w.f.) and a cationic retarder (1-2% o.w.f.) are added as leveling agents. A small amount (0.3-0.5% o.w.f.) of a lubricating agent is also added to prevent excessive friction of the fabric during its movement in the dyeing machine. The fabric is inserted and the machine is run for 5 min. Then the dissolved cationic dyes are added slowly over a period of 10 min., after which the temperature is raised at 2-3°F/min. to just below the glass transition temp. (anywhere between 68°C and 85°C) (155°F and 185°F) of the fibers. Dyeing temperature is kept at this point for 15 min., after which it is raised to the boil at 1°F/min. Then the dyeing continues at the boil for about one hour. If shade correction is needed the dye-bath must be cooled to below the glass transition temp. of the fibers before the additions are made. When the proper shade is obtained, the dye-bath is cooled very slowly to about 10°C (18°F) below the glass transition temp.(to avoid the formation of creases), and then dropped. The dyeing is completed by rinsing in the regular manner.

Dyeing time at maximum temperature can be reduced significantly if dyeing is carried out under pressure at 104°C-107°C (220-225°F). Also, when dyeing under pressure, level results are easier to obtain. However, maximum dyeing temperatures must not exceed 110°C (230°F) to avoid possible damage to the fibers.

16.4 Factors Influencing the Dyeing with Cationic Dyes

When dyeing acrylics with cationic dyes, the dyer should have the proper information regarding the fibers and the dyes that are chosen [1, 2, 3].

Fiber Characteristics

Acrylic fibers from different sources may differ from each other greatly in their glass-transition temperature, degree of porosity (ease of penetration), and their saturation value.

The degree of porosity and the glass transition temperature affect the rate of dyeing. Wet spun acrylic fibers have more open structures and lower glass transition temperatures than those of dry spun acrylic fibers; therefore, the wet spun fibers dye at a faster rate. Other factors can also affect the dyeability of the fibers such as variations in the degree of drawing of the fibers.

The saturation value indicates the number of available dye-sites (anionic groups). When the number of dye-sites is limited, the dye that dyes at a faster rate may block the other dyes in the dye-bath from entering the fibers. The saturation value for a given fiber is constant, but the maximum amount of a specific dye that can be bonded to the fibers varies from dye to dye; the higher the molecular weight of a dye the higher the amount of dye (by weight) that can be bonded to the fiber.

Characteristics of Cationic Dyes

The proper choice of cationic dyes to be used in a combination, is the single most important decision the dyer has to make. Since cationic dyes are very difficult to migrate, and the number of available dye-sites is limited, dyes used in a combination must exhaust at a similar rate. Accordingly the cationic dyes, which vary greatly in their substantivity, are classified according to their rate of exhaustion.

A subclassification of the cationic dyes based on their exhaustion behavior, has been developed by Bayer Co. (now Mobay Chem Corp). In this subclassification, each dye is assigned a K-value (combination constant) between 1 and 5 in increments of 0.5. Cationic dyes with a K-value of 1 are the fastest to exhaust, those with a K-value of 5 have the slowest rates of exhaustion. In order for dyes that are used in a combination to exhaust on tone they should have the same or similar K-values, not to exceed a difference of more than 1 unit.

When dyeing dark shades, dyes with K-values of 3 and less are preferred since these dyes have a high rate of exhaustion and dyeing can be achieved in a reasonable amount of time. For light shades, dyes with K-values of 4-5 should be used. Dyes with low K-values will strike too fast and may lead to unlevel dyeing.

16.5 Retarders for Cationic Dyes

Various cationic and anionic retarders are offered by manufacturers for slowing down the rate of exhaustion. In addition, sodium sulfate has been found to act as a mild retardant, and is used often in combination with the other retarders.

Cationic Retarders

The retarders most often used in dyeing acrylics with cationic dyes are the cationic retarders. These retarders are cationic organic compounds (e.g. quaternary amines with a long hydrocarbon chain), soluble in water, that can be described as 'colorless cationic' dyes. The cationic retarders compete with the cationic dyes for the anionic dye-sites, and occupy them temporarily.

The main limitation of the cationic retarders is that they may form strong bonds with the fibers and then block some of the anionic dye-sites permanently, thus reducing significantly the number of dye-sites [4]. K-values are also assigned to the cationic retarders which should be selected to have a similar or slightly lower K-value than that of the dyes. When the K-value of the retarder is much lower than that of the dyes, it may block them, and if its K-value is higher than that of the dyes it may be ineffective in slowing down the rate of exhaustion of the dyes.

Cationic Polymeric Retarders

These cationic substances, which may contain more than a hundred cationic groups on each polymeric molecule, adhere to the fibers' surface but cannot diffuse into them. They act by forming a positive coating on the fibers which prevents the dye-cations from diffusing rapidly [5].

Anionic Retarders

The anionic retarders are water soluble organic compounds, anionic in nature. They form complex salts with cationic dyes, thus reducing the amount of dye-molecules in the dye-bath available for exhaustion. With increasing dyeing temperature the complexes gradually dissociate and free dye-cations become available for exhaustion.

The main disadvantage of the anionic retarders is that the complexes formed between them and the cations of the dyes may be too stable. Therefore, a considerable amount of the dye will be left in the dye-bath unused. In addition, the complexes formed may sometimes precipitate out. To avoid precipitation, usually a non-ionic surfactant is included in the dye-bath.

References

1. Emsermann Hubert, Dyeing of Acrylic Fibers with Cationic Dyestuffs, Mobay Chemical Corp., A.A.T.C.C., Jan. 11, 1990
2. Herlant Michel A., Crompton & Knowles Corp., Dyeing of Acrylics, A.A.T.C.C. Workshop, Jan. 11, 1984
3. Phillips, R. R., Mobay Chemical Corp., Rock Hill, South Carolina, Jan. 1974
4. Landerl & Baer, Am. Dyestuff. Rep. 54, 1965, 222
5. Dullaghan & Ultee, T.R.J. 43, 1973.

Chapter 17.

DYEING POLYESTER/COTTON AND OTHER BLENDS

17.1 Union and Cross Dyeing

In union dyeing, the components of a blend are dyed the same shade to obtain a solid color. This method is often applied to polyester/cotton intimate blends (fiber blends). Union dyeing can be applied from a single dye-bath or by dyeing components of the blend one at a time.

The method of obtaining a.multi-color effect on a blend is referred to as cross-dyeing. When two or more contrasting colors are desired, the fibers in the blend must have varying affinities to dyes.

The main problem in applying contrasting colors is the staining of one of the fiber types in the blend with dyes used for the other fiber type. Staining in union dyeing is less critical since all fibers are dyed the same shade. However, in both cases the unfixed staining dye may lead to poor wash-fastness.

17.2 Exhaust Dyeing of Polyester/Cotton Blends

The preparation of the cotton in the blend is carried out in a manner similar to that of 100% cotton. In general, polyester fibers can withstand the preparation treatments that cotton fibers undergo. However, caustic scouring at the boil for prolonged durations can cause damage to polyester. Therefore, milder conditions for the removal of cotton wax are employed.

There are several methods of dyeing polyester/cellulose blends, each with its own advantages and drawbacks. The polyester is dyed with disperse dyes, and the cotton is dyed by any dyes suitable for cellulosic fibers.

The following dyeing procedures are examples of common methods for union dyeing of polyester/cotton blends with disperse and reactive dyes.

The Two-Bath Method
In this method, the polyester is dyed first. After the dyeing of the polyester, whenever needed, a reduction-clearing can be applied to remove unfixed disperse dye. The cotton is then dyed by the selected dyes suitable for dyeing cellulose.

Best results are obtained by this method, but it is also the most expensive one. This method allows the use a reduction-clearing, so that any disperse dyes that stain the cotton or adhere to the surface of the polyester fibers can be removed. As for the disperse dyes inside the polyester, they will not be affected under the conditions used in the reduction-clearing, and the dyeing the cotton. This method can also be used for cross-dyeing of polyester/cotton, when using the proper disperse dyes (causing little or no staining of the cotton).

The One-Bath Two Step Method (With disperse and reactive dyes)
The dye-bath is prepared with the disperse dyes and the proper dyeing assistants. The temperature is raised to the dyeing temperature of the polyester. After dyeing the polyester, the dye-bath is cooled to the dyeing temperature of the cotton. Then the reactive dyes are added and the dyeing of the cotton is carried out. This method cannot include a reduction-clearing after dyeing since the chemicals used for the reduction will destroy the reactive dyes. The use of this method saves time and energy.

Sometimes it is also possible to dye from a single dye-bath to which both dye types are added simultaneously together with the dyeing assistants. By control of bath temperature and pH, conditions are established to first introduce the disperse dye into the polyester, then bond the reactive dye to the cellulosic fiber.

The Reverse Method-Two Bath Dyeing (with disperse and reactive dyes)
In this method the cotton is dyed first with reactive dyes, but the lengthy after-scouring is omitted. Then the dyeing of the polyester takes place as usual. The advantage here is that the dyeing of the polyester serves as the after-scouring for the reactive dyes. However, the selected reactive dyes for this method must withstand the conditions used in the dyeing of the polyester.

17.3 Continuous Dyeing of Polyester/Cotton

In continuous dyeing of polyester/cotton blends, the polyester is dyed with disperse dyes, and the cotton in the blend is dyed with one of the following dye types: vat, reactive, or sulfur dyes. A common continuous range for dyeing polyester/cotton blends is shown in Fig. 5.10. In this range the following steps occur:
dye-pad/pre-dry/dry/thermosol/cool/chemical-pad/steam/wash, etc.

Since the fabric is treated at high speeds (100 yards per min. and more), a faulty dyeing may be detected only after running hundreds of yards. Accordingly, the individual units are carefully controlled to insure uniformity. The two padding machines in the range are so controlled that the same pressure is applied along the full width of the fabric. In drying and curing ovens, the temperature is controlled to be the same along the full width of the fabric. This will prevent variations from selvage to selvage and from selvage to center. In addition, the dye-bath content should be constantly monitored, as the concentration of dyes and/or dyeing assistants may change.

When dyeing the cotton portion with sulfur dyes, soluble sulfur dyes are used, and they are applied from the chemical pad along with additional amounts of a base and a proper reducing agent.

17.4 Continuous Dyeing Procedures

Continuous Dyeing of Polyester/Cotton with Disperse and Vat Dyes
(thermosol/pad/steam)

In this application, both dyes are applied from the same dye-bath (dye-pad). The polyester is dyed with the disperse dyes by the thermosol method, and the cotton is dyed with the vat dyes by the continuous pad/dry/pad/cure method. These two procedures are combined and the application consists of the following steps:

1. <u>Dye-pad.</u> The padding bath contains the following :

        vat dyes (IN type)
        disperse dyes (high energy type)
        acetic acid to pH 4.5 - 5
        anti-migrant
        dispersing agent (if necessary)
        wetting agent (0-5 g/l)

Remarks.

-Since the vat dyes are hydrophobic they will tend to diffuse into the polyester during the thermosol stage as do the disperse dyes. To avoid this, IN vat dyes or other vat dyes of a relatively large molecular size are used. These dyes are of a much larger molecular size compared to that of disperse dyes.

-The disperse dyes are of the high energy type. The preferred ones are those that actually sublime at the thermosol stage to allow movements of disperse dyes from the surface of the cotton in the blend into the polyester fibers. For example, when padding a 50/50 polyester/cotton blend, most of the dye solution is absorbed onto the cotton surface. During thermosoling, the disperse dye moves from the cotton to the polyester by contact migration or sublimation.

-Acetic acid is used to ensure an acidic pH which is needed to protect many disperse dyes that are sensitive to a pH above 7 at high temperatures.

-An anti-migrant, usually a thickening agent such as sodium lignate, is used to prevent the movement of dyes from wet areas to dry ones during the drying stage. The dyes also tend to move to the surface by the capillary movement of the water toward the fabric surface. This can also lead to face-to-back shading or crocking problems.

-The amounts of wetting agent and dispersing agent used should be kept at a minimum. The weak attractions between disperse dyes and wetting or dispersing agents, may slow down the movement of dyes into the polyester, during the thermosol stage. Therefore, excessive amounts of wetting and dispersion agents in the dye-bath may decrease the yield of dyeing.

2. Pre-Dry (see Chapter 5)
3. Dry (see Chapter 5)
4. Thermosol (see 14.8)
5. Cooling Cans

Remarks.

-A fast cooling is accomplished by passing the fabric over cooling cans. It will prevent the hot fabric, coming out of the oven, from gradually heating the chemical pad that follows.

-The cooling step also provides time for the bone-dry cotton to gain some moisture.

6. Chemical Pad

The chemical pad contains the following:

a reducing agent (e.g. hydrosulphite)

sodium hydroxide

a sequestering agent

a dispersion agent and/or wetting agent if needed.

Remark
-A significant advantage of this method is that the reduction treatment of the vat dyes also serves as a reduction-clearing for the unfixed disperse dye.

7. Steam

After-treatments in Wash Boxes:

8. Rinse (neutralize)   9. Oxidize   10. Soap at the Boil   11. Rinse

Dyeing Polyester/Cotton with Disperse and Reactive Dyes
(thermosol/pad/steam)

In this application, the same common continuous range described above can be used (Fig. 5.10.). The chemical pad contains the disperse and reactive dyes. In addition, an anti-migrant, a dispersing and/or a wetting agent, and acetic acid are added as described above. After padding, drying, thermosoling, and cooling, the fabric is passed through the chemical-pad. The chemical pad contains the alkali for the fixation of the reactive dyes, and common salt. The presence of salt in the chemical-pad prevents the reactive dyes on the fabric from contaminating the chemical-pad

Note that a reduction-clearing is not taking place here as in the application with disperse and vat dyes. A reduction-clearing cannot be applied here as a separate step as it will destroy the reactive dyes on the cotton.

17.5   Dyeing Wool Blends

Dyeing Nylon / Wool

Both fibers can be dyed with the same acid dyes from a single dye-bath to obtain solid shades. Using disperse dyes for dyeing the nylon in the blend, even for light shades, is not recommended since disperse dyes stain wool heavily.

Under the same pH, acid dyes are more substantive to nylon than wool. Therefore, when dyeing light shades with acid dyes, the nylon in the blend will tend to dye deeper. However, for darker shades the opposite occurs. The saturation point of acid dyes on nylon is much lower than on wool. Past the saturation point of nylon, the wool will dye much deeper, and therefore this method is limited to light to medium shades. Dyes selected for a particular dyeing should exhaust to a similar extent on both fibers in the blend.

When dyeing light shades, retarders are used to slow the exhaustion on nylon which would other wise dye at a faster rate. These retarders are anionic compounds such as (colorless acid dyes) which temporarily occupy amino groups on the nylon fibers. An example is: an alkyl-benzene sulfonate.

Dyeing Polyester/Wool

The polyester is dyed with disperse dyes, and the wool with acid dyes. Since the wool cannot be exposed to very high temperatures as in the high pressure dyeing of polyester, dyeing is carried out at or a few degrees above the boil. Therefore, dye-carriers have to be used to facilitate the dyeing of the polyester.

The main problem with dyeing polyester/wool blends is that disperse dyes tend to heavily stain wool. The preferred disperse dyes are those that have a low tendency to stain the wool, or can be easily removed from it by after-scouring. Dye-carriers should also be carefully selected since they have a direct effect on the degree of wool staining. At the beginning of dyeing, below the boil, the wool is heavily stained, and as the dyeing temperature approaches the boil the disperse dyes move slowly to the polyester. Accordingly increasing temperature and time of dyeing decreases the amount of disperse dye left on the wool fibers. However, prolonged dyeing time may cause damage to wool. The reduction action of wool should also be taken into consideration when choosing proper disperse dyes.

When dyeing the blend by a two-bath method, the polyester is dyed first. The dye-bath is prepared with the disperse dyes and a dye carrier(s), and the pH of the bath is adjusted with acetic acid and ammonium acetate to pH 5-6. The temperature is raised to the boil or slightly above the boil (105ºC) (220ºF), and dyeing continues at maximum temperature for 45 to 60 min. A reduction-clearing is applied at this stage to remove unfixed disperse dyes. The reduction clearing is carried out with hydrosulphite (3g/l) ammonia (3g/l), and a nonionic detergent (1g/l), at 50-55ºC (122-131ºF) for 20 min. These mild conditions are necessary since wool is sensitive to alkali. After dyeing the polyester, the wool is dyed in the regular manner with acid dyes suitable for the normal use of the garment.

Note that even after reduction-clearing, disperse dye migration from polyester to wool may occur during the dyeing of the wool.

Dyeing from a single bath is recommended for light to medium shades. The dye-bath contains the disperse and acid dyes, a dye-carrier, a dispersing agent, ammonium acetate, and acetic acid to pH 5-6. The temperature is raised slowly to the boil over 45 min. Dyeing is continued at 100-105ºC (212-221ºF), but not higher, for 60 min. After dyeing, the material is washed with a nonionic detergent at 60ºC (140ºF).

Dyeing Acrylic/Wool Blends

The acrylic component is dyed with cationic dyes and the wool component with acid dyes. Disperse dyes are not used as they heavily stain wool. The acrylic/wool blend can be dyed from a single bath but this method is only recommended for light to medium shades. A two-bath method is also preferred when applying contrasting colors (cross-dyeing).

When dyeing deep shades, the wool is heavily stained by the excess of the cationic dyes used. Therefore, the more expensive two-bath method is employed, which allows the clearing of cationic dyes from the wool. In this method the acrylic component is dyed first. A reduction clearing is then applied, and it is carried out under acidic conditions with the proper reducing agent (e.g., zinc formaldehyde sulphoxylate). Then the wool is dyed with any of the acid or metallized dyes.

When dyeing both fibers from a single dye-bath, cationic and acid dyes can be applied together. However, the colored ions may form insoluble complex salts. In order to prevent precipitation, nonionic compounds are used as stabilizers. The stability of the dye-bath for any particular combination should be carefully checked. In this method the dye-bath is prepared as follows:

set the bath at 120°F and add:

a) acid dyes

b) acetic acid to pH 4-5.

c) A retarding agent for the cationic dyes

d) anti-precipitating agent (non-ionic)

e) cationic dyes

The dyeing is then carried out following the same procedure as in the dyeing of 100% acrylic fibers with cationic dyes. Both fibers are dyed at the same time. Note that the attraction of the colored ions to each other has a retarding effect on their rate of exhaustion. A small amount sodium sulfate (less than 10% o.w.f.) can be used, and it will act as a retarder for both dyes. A pH of 4-5 in the dye-bath is preferred since this pH is within the range of the iso-electric point of wool, and is also the proper pH for the application of the cationic dyes. The selected acid dyes are those that exhaust well at a pH of 4-5.

References

1. Otis E. Melton, Solving Problems in Continuous Dyeing, American Dyestuff Reporter, October 1976, p. 59.
2. Claude S. Hughey, Continuous Dyeing of Polyester/Cotton Blends By the Thermosol/Pad-Steam Process, American Dyestuff Reporter, September 1979, p. 42.
3. Lamar R. Smith and Otis E. Melton, Troubleshooting in Continuous Dyeing, Text. Chem and Col., May 1982, p. 38.
4. Brent Smith, and Leon Moser, Troubleshooting in Dyeing- : Continuous Dyeing, A.D.R., May 1987, p. 36.
5. J. Park and S.S. Smith, A Practical Introduction to the Continuous Dyeing of Woven Fabrics, Roaches (Engineering) Limited, 1990, England
6. The Dyeing of Synthetic-polymer and Acetate Fibers, Edited by D.M. Nunn, Chapter 7. Dyeing of Blends, By J. Shore, 1979 Dyers' Company Publications Trust, West Yorkshire, England.
7. Wool, A Sandoz Manual, Sandoz Ltd, Basel Switzerland, 15. Dyeing of Polyester/Wool Blends, 16. Dyeing of Wool/Acrylic Blends.
8. Dyeing and Finishing of Polyester Fibres, A BASF Manual, Dyeing Polyester/Cotton, Polyester/Wool, and other blends with Polyester.
9. Dyehouse Manual for the Shift Dyer, VPI Marketing, Inc., Albemarle, N.C., 1981.

APPENDIX

A. General Precautions

For shade reproducibility, all conditions used in a specific application must be recorded accurately and followed precisely in future runs. These conditions include temperatures, time, concentrations of dyes and dyeing assistants, liquor ratio, source of dyes and chemicals, etc.

All dyeing assistants in any dye-bath should be used only if necessary.

Avoid using excessive amounts (larger than recommended) of any additive.

All dyes and dyeing assistants should be added to the dye-bath in the form of solutions or dispersions. Concentrated dye solutions or dispersions should be filtered to prevent undissolved or undispersed dye particles from entering the system.

Prolonged movements of delicate fabrics during long dyeing cycles, may cause pilling, abrasion, and/or rope marks.

Dyeing at temperatures higher than those recommended may cause damage to fibers and/or dyes.

All chemicals and dyes left unused should be disposed in a manner consistent with OSHA (Occupation Safety and Health Administration) and EPA (Environmental Protection Administration) regulations.

B. Dye-Selection

The proper choice of dyes is the single most important decision the dyer has to make. Not only the class of dyes has to be decided, but also within that class the appropriate subclass of dyes should be carefully selected.

Be aware that different results may be obtained by using the same dye from a different manufacturer. Commercial products of the same dye can differ in degree of purity, particle size, type and amount of fillers mixed with the dye, etc.

When choosing dyes for a specific dyeing the following should be taken into consideration:

Main Considerations

Suitability of the dyes for the type of fibers to be dyed and Method of Application chosen (e.g. atmospheric, under pressure, etc.)

Fastness requirements (end use)

Leveling properties (ease of penetration and migration properties)

Shade requirements (light or dark shades) and Color Matching (color consistency under different sources of light)Compatibility of the dyes in the bath with each other and with dyeing assistants

Cost, Availability, and Inventory Control

Other Considerations

Physical form (powder or granular or liquid, and particle size)

Solubility or Dispersion stability

Staining of other fibers in the substrate

Coverage of barre and other irregularities in the substrate

C. Possible Causes for Inappropriate or Faulty Dyeing

IMPROPER PREPARATION
-uneven heat treatments
-insufficient removal of impurities
-deposits of insoluble materials
-uneven absorbancy
-uneven mercerization
-inadequate moisture content
-variations in substrate structure (e.g. barre, dead cotton, or other damages in fibers)
-residues of chemicals such as bleaching agents, bases, etc.

VARIATIONS IN THE WATER CONTENT
-degree of hardness
-presence of transition metals
-variations in pH
-chlorine, etc.
-other contaminations

ERRORS IN DYE-BATH PREPARATION
-mistakes in weighing dyes, dyeing assistants, and the substrate
-variations in liquor ratio
-improper preparation (filtration) of the concentrated dye solutions
-new source of purchasing the dyes and/or the dyeing assistants

DYEING EQUIPMENT
-defects in the equipment causing loss of dye liquor during application
-failure of control(s), e.g. temperature and/or time controls, and change in speed of dye-bath circulation and/or circulation of the substrate.

D. Fastness Requirements

In normal use, the most desirable properties for the dye to have are:
1. wash-fastness
2. light-fastness
3. fastness to rubbing (crocking).

Also, for proper performance during applications subsequent to dyeing (e.g. heat-setting, pleating, pressing, etc.), fastness to thermal treatments is usually required.

Other fastness properties often required include fastness to:
dry-cleaning or solvent bleeding,
atmospheric contaminants such as nitrogen oxides,
perspiration,
steam,
and various bleaching agents.

Except for light-fastness, fastness ratings are given by using the scale of 1 to 5, where:

5 =	excellent
4-5 =	very good
4 =	good
3 =	fair
2 =	poor
1 =	very poor

For light-fastness a scale of 1 to 8 is used where:

8 =	outstanding
7 =	excellent
6 =	very good
5 =	good
4 =	fairly good
3 =	fair
2 =	poor
1 =	very poor

In reacting with light the dye can loose its color or change its hue. A reduction in depth of shade is not as objectionable as a change in the hue of the dye.

E. Relative Size of Dye Molecules

Type of Dyes	Relative Size
Acid leveling	small - medium
Acid milling	medium - large
1:1 metal complex	small - medium
2:1 metal complex	large - very large
Azoic	small
Neutral dyeing acid dyes	large - very large
Basic	small - medium
Direct	very large
Disperse	small
Reactive	small - medium
Sulfur	very large
Vat	medium to very large

F. Suitability of Dyes for Different Fibers

CLASS OF DYES	Cellulose (cotton)	Protein (wool)	Nylon	Polyester	Acrylic
Acid		√√	√√		√
Azoic	√√		√	√	
Basic			√	√	√√
Direct	√√	√	√		
Disperse			√	√√	√
Reactive	√√	√	√		
Sulfur	√√				
Vat	√√				

√ - used to a small extent, or for light shades only   √√ - most commonly used

G. The Ionization of Water and The pH Scale

pH	[H+] mole/liter	[OH-] mole/liter	p[OH⁻]	
0	1	10^{-14}	14	
1	10^{-1}	10^{-13}	13	strongly acidic
2	10^{-2}	10^{-12}	12	
3	10^{-3}	10^{-11}	11	moderately acidic
4	10^{-4}	10^{-10}	10	
5	10^{-5}	10^{-9}	9	slightly acidic
6	10^{-6}	10^{-8}	8	
7	10^{-7}	10^{-7}	7	neutral (pure water)
8	10^{-8}	10^{-6}	6	
9	10^{-9}	10^{-5}	5	slightly basic
10	10^{-10}	10^{-4}	4	
11	10^{-11}	10^{-3}	3	moderately basic
12	10^{-12}	10^{-2}	2	
13	10^{-13}	10^{-1}	1	strongly basic
14	10^{-13}	10^{-1}	0	

(pH decreases with increasing hydrogen ion concentration)

H. pH Values of Chemicals Used in Wet Processing of Textiles

compound	Formula	Conc. molar.	Conc. %w/w	pH (approx.)
Hydrochloric acid (muriatic acid)	HCl	0.1	0.37	1.1
"		0.01	0.04	2.0
"		0.001	0.004	3.0
Sulfuric acid	H_2SO_4	0.5	4.9	0.3
"		0.05	0.49	1.2
"		0.005	0.05	2.1
Phosphoric acid	H_3PO_4	0.1	1.0	1.5
Formic acid	HCOOH	0.1	0.46	2.3
Acetic acid (vinegar)	CH_3COOH	1.0	6.0	2.4
"		0.1	0.6	2.9
"		0.01	0.06	3.4
"		0.001	0.006	3.9
Carbonic acid (carbon dioxide + water) saturated				3.8
Ammonium chloride	NH_4Cl	0.1	0.53	5.3
Sodium chloride (common salt)	NaCl			7
Sodium sulfate (glauber salt)	Na_2SO_4			7
Sodium bicarbonate (baking soda)	$NaHCO_3$	1.0	8.4	8.4
"		0.1	0.84	8.4
"		0.01	0.084	8.4
Sodium acetate	CH_3COONa			9.9
Ammonium hydroxide (ammonia)	NH_4OH	0.01	0.04	10.7
"		0.1	0.35	11.1
"		1.0	3.5	11.7
Sodium carbonate (soda ash)	Na_2CO_3	0.01	0.11	11
"		0.05	0.55	11.5
"		0.1	1.1	11.8
Sodium meta silicate	$Na_2SiO_3 \cdot 5H_2O$		0.05	11.4
"			0.1	11.5
"			1.0	12.4
Sodium hydroxide (caustic soda)	NaOH	0.01	0.04	12.1
"		0.1	0.4	13

I. Temperature Conversion
Degrees Fahrenheit to Degrees Centigrade

°F	°C	°F	°C
70	21	194	90
75	24	198	92
79	26	203	95
84	29	208	98
90	32	212	100
95	35	221	105
100	38	230	110
104	40	239	115
111	44	248	120
115	46	257	125
120	49	266	130
124	51	275	135
129	54	284	140
133	56	293	145
138	59	302	150
140	60	311	155
145	63	320	160
149	65	329	165
154	68	338	170
158	70	347	175
160	71	356	180
165	74	365	185
167	75	374	190
172	78	383	195
176	80	392	200
178	81	401	205
183	84	410	210
185	85	419	215
187	86	428	220

REVIEW QUESTIONS

Introduction

1. The bleeding of color out of a cloth during laundering could be due to two different reasons. What are they?
2. Name five prime requisites for a chemical to serve as a useful dye.
3. What do the following terms describe:
   a. substantivity   b. exhaustion   c. after-treatment
   d. strike   e. leveling agent   f. diffusion
   g. migration   h. retarder
4. Name three advantages of dyeing raw stock over dyeing piece goods.
5. Why do dyes with high substantivity have poor migration?
6. How does the rate of circulation of the dye-liquor through the material affect the rate of dyeing?
7. What are the advantages and disadvantages of dyeing at low liquor ratios? (Chapter 5)
8. What happens to thermoplastic fibers when they are heated to their glass-transition temperature?
9. How is the contact between the fibers and the dye solution achieved in the following dyeing machines: (Chapter 5)
   a. jig   b. package dyeing   c. jet   d. beck
10. Of the following dyeing machines which are the two more suitable ones for dyeing acetate fabrics: (Chapter 5)
    a. beck   b. jig   c. beam   d. jet. Explain

Detergents and Scouring

1. What is soap?
2. Why is soap classified as an anionic detergent?
3. What are the limitations of soap?
4. Why is soap insoluble in acidic solutions?
5. Why is soap insoluble in hard water?
6. What type of a pH has a soap solution? Why?
7. Why is the sodium salt of a fatty acid more soluble than the fatty acid itself?
8. Soaps are made of hydrocarbon chains with 12 to 18 carbons. What will be the disadvantage of a chain:
   a. with less than 12 carbons?   b. with more than 18 carbons?
9. Is ethanol as good a cleaning agent as soap? Explain. Consider both the hydrophobic and the hydrophilic parts of the molecule of ethanol.
10. What is the advantage of using an ionic synthetic detergent instead of soap?
11. What is the difference between an anionic syndet and a cationic syndet?
12. What type of detergents are sodium alkyl-sulphate and sodium alkyl-sulphonate? How are they affected by hard water? Explain.
13. What type of a pH has a solution of sodium alkyl-sulphate? Why?

14. What are the two things that can occur when mixing anionic and cationic surfactants in the same dye-bath?
15. Why are cationic surfactants not used in the scouring and preparation of fibers for dyeing and finishing? What are they used for?
16. Sodium carbonate (Na_2CO_3) is often added to a scouring bath whether soap or a synthetic detergent is used as the scouring agent. Why? (Give two reasons)
17. Describe a micelle made of an oil particle and an anionic syndet.
18. When scouring wool, why are nonionic detergents preferred over soap?
29. How do sequestering agents remove hardness from water?
20. Explain why is the complex made of EDTA and Ca^{++} or Mg^{++} water soluble?
21. Two types of metals can interfere with dyeing procedures. What are they? How are the complexes between these metals and EDTA affected by lowering the pH of the 22 dye-bath?
23. Explain why scouring of wool is performed on its loose fibers form while scouring of cotton is delayed until shortly before dyeing.

Classification of dyes and Fibers

1. What are the differences between a dye and a pigment? Consider:
   a. solubility   b. substantivity   c. fastness properties
   d. molecular size   e. intensity of color
2. Is it possible to use pigments in exhaust dyeing?
3. Compare the solubility of ionic dyes with that of disperse dyes. Are the disperse dyes water soluble? Explain.
4. List the following fibers in order of increasing moisture content:
   cotton, rayon, polyester, nylon, polypropylene, wool, and acrylic fibers. Explain why is the moisture regain of nylon is ten times more than that of polyester?
5. Indicate whether the following fibers are hydrophilic or hydrophobic, and explain in each case why?
   a. cotton   b. polyester   c. acetate   d. wool
   e. acrylic   f. polypropylene
6. Name three fibers that can be dyed with disperse dyes and three fibers that can be dyed with soluble dyes. Explain why?
7. How does the molecular size of a pigment affects its performance?

Preparatory Operations

1. When enzymes are used for desizing, what is the chemical change that starch undergoes?
2. How is polyvinyl alcohol (PVA) removed when used as a sizing agent?
3. What is the main advantage of using partially hydrolyzed PVA as a sizing agent? Explain.
4. What is the chemical nature of a wetting agent as compared to detergents?
5. Why is a detergent not required in caustic scouring of cotton?
6. A sequestering agent is added to the caustic-scouring bath. Explain why?
7. Explain how are the following substances removed from cotton during caustic scouring:

  a. cotton-wax   b. fats   c. proteins   d. pectins
8. Why is caustic scouring applied to cotton before bleaching?
9. What are the bleaching agent(s) that you would prefer to use on each of the following fibers:
  a. wool   b. cotton   c. nylon   d. acrylic   Explain why?
10. What is the chemical reaction that takes place when bleaching with hydrogen peroxide?
11. What are the advantages of using hydrogen peroxide over sodium hypochlorite for bleaching textile fibers?
12. When bleaching with hydrogen peroxide, what is the purpose of using sodium silicate? (Give three advantages). What are the disadvantages of using sodium silicate?
13. Explain why the reactivity of H_2O_2 increases with increasing the pH of its solution in water?
14. Certain metal ions accelerate the decomposition of H_2O_2. Other metal ions are effective stabilizers for H_2O_2. Give examples for the two types of metals. How are these metals affected by the presence of sodium silicate?
15. What are the advantages of mercerizing under tension?
16. Why does improper mercerization cause unlevel dyeing?
17. What are the main reasons for mercerizing nowadays?
18. Explain how optical brightening agents make a yellow fabric to appear white?
19. The most common optical brighteners are derivatives of stilbene. Explain why are these brighteners sensitive to sunlight and to bleaching agents? (examine the chemical structure of stilbene).
20. How are optical brighteners applied to cotton?
21. Why is a proper and uniform preparation more important for continuous dyeing than for exhaust dyeing?

Color and Chemical Constitution

1. What do the following terms describe:
  a. a chromophore   b. an auxochrome   c. a chromogen
2. What part of the visible light is absorbed when the color of the object seen at day light is:
  a. magenta   b. red   c. brown   d. grey   e. cyan
3. By mixing yellow, cyan, and magenta paints, how would you make:
  a. blue   b. red   c. olive   d. navy blue   e. brown   grey
4. A red light, a green light, and a blue light, are alternately projected at the same spot on a white screen. What color will appear on the screen when:
  a. the red light and the blue light are projected simultaneously?
  b. the green light and the red light are projected simultaneously?
  c. the blue light and the green light are projected simultaneously?
5. What color would a blue car appear to be under an all-yellow sodium vapor street light? Explain.
6. Samples of the same fabric were dyed in the same shade, using different dyes for each. Must their shades always appear the same under different sources of light?
7. Explain why graphite (carbon black):
  a. absorbs the entire visible light?
  b. is a good electrical conductor?
  c. is planar?

d. is used extensively in mass coloration (solution dyeing)? Give two reasons.
8. Arrange the following dyes in order of: a. increasing molar absorptivity (intensity of absorption), b. increasing fastness to light:
   a. azo dyes   b. triarylmethane dyes   c. anthraquinone dyes
   d. phthalocyanine dyes.
   How do light fastness and intensity of absorption relate to each other?
9. Why do azo red dyes have a lower molecular weight as compared to azo blue dyes?

Dyeing Wool with Acid Dyes

1. Why are acid dyes classified as anionic dyes?
2. What are the components of an acid dye? Consider:
   a. color   b. solubility   c. bonding group(s)
3. What are the structural features in and acid dye that affect its wash-fastness?
4. Why do acids promote exhaustion of acid dyes?
5. Explain why acid dyes combine directly with animal fibers but not with vegetable fibers?
6. How does sodium sulfate acts as a leveling agent for acid dyes?
7. How is the dyeing of wool with acid dyes controlled?
8. Explain how sodium sulfate competes with acid dyes for the dye sites on wool?
9. What is the advantage in using neutral-dyeing acid dyes, and why is ammonium sulfate added to the dye-bath?
10. When dyeing with neutral-dyeing acid dyes, why is diammonium-phosphate added? Why do we not use acids in the application?
11. Why is it difficult to obtain level dyeing with neutral-dyeing acid dyes? (give two reasons)
12. Why are the neutral-dyeing acid dyes more durable to laundering then the acid-leveling dyes? Give two reasons.
13. When dyeing with neutral-dyeing acid dyes, the amount of sodium sulfate, if used at all, should not exceed 5% (o.w.f.). Why? (see: effect of salt on dye-solubility)
14. What is the advantage of dyeing wool by the top chrome method? What are the disadvantages of this method?
15. Why does an after-treatment with certain chrome compounds increase the durability of some acid dyes to laundering?
16. Two types of metallized dyes for wool are noted. Explain the difference in their application. How can the difference in their behavior be explained?
17. When dyeing wool, what type of dyes would you consider first if you were most concerned with:
    a. fastness to milling   b. level dyeing   c. rapid dyeing
    d. bright colors   e. durability to laundering and light
18. Choose a structure of an acid dye and indicate in it the following:
    a. an electron donor   b. an electron acceptor
    c. a chromophore   d. a chromogen   e. an auxochrome
    f. a solubilizing group   g. a bonding group
19. What is the oxidation state of chrome when applied to the fibers, and what is its final oxidation state when bonded to the fibers? How does the pH affect the stability of the chrome in its different compounds?
20. Why can ammonia or sodium carbonate in small amounts partially strip acid dyes from wool?

21. Why is sulfuric acid sometimes added toward the end of the dyeing that has been started with acetic acid?
22. What is the iso electric point of wool? Why is it preferred to dye wool as close as possible to its iso electric point? (consider the ionic cross-links of wool)
23. How are neutral dyeing acid dyes modified by the dye manufacturer to be used at the iso electric point of wool?
24. Scouring wool with natural soap prior to dyeing is not recommended. Explain why?
25. Compare the method of achieving level dyeing through migration (in conventional exhaust dyeing) with the method of achieving level dyeing through controlled exhaustion. Consider:
    a. the nature of the dyes recommended for each method
    b. dyeing temperature
    c. controlling dyeing conditions
    d. reproducibility of the required shade
    e. dyeing deep (dark) shades

Dyeing Cellulose with Direct Dyes

1. What do acid and direct dyes have in common?
2. What makes the direct dyes substantive to cellulose?
3. How do direct dyes bond to cellulosic fibers? (consider two types of bonds).
4. Explain why the addition of salt, during the application of direct dyes to cellulosic fibers, promotes exhaustion?
5. Why are direct dyes on cotton less durable to washings than any of the other dyes used for cellulosic fibers?
6. What is the advantage of dyeing with direct dyes over the other methods of dyeing cellulose?
7. Why do cationic fixing agents improve the wash-fastness of direct dyes?
8. What features does a dye molecule need to have in order to be used as a direct dye?
9. How is the dyeing of cotton with direct dyes controlled? Consider the sub groups of direct dyes: A, B, and C.
10. In a typical application, direct dyes with maximum absorption temperature in the range of 160°F were used. However, the actual dyeing time mostly took place at the boil. Explain why?
11. Explain why many direct dyes are used as neutral-dyeing acid dyes for polyamides?
12. Can a neutral-dyeing acid dye be used as a direct dye for cotton? Explain.
13. How do you strip a direct dye, that was after-treated with copper sulfate, from cellulosic fibers?
14. Explain why a direct dye with a bright green color is expensive?
15. What type of direct dyes are sold in a more concentrated form with a minimum amount of electrolytes as diluents? (consider the sub groups A, B, and C.) Explain.
16. In union dyeing of cotton/wool blends with direct dyes, the wool tends to dye deeper. Explain why?
17. When dyeing with direct dyes, what causes the problem of 'late exhaustion'? What should be done to avoid missing the correct shade because of 'late exhaustion'?
18. Sequestering agents must not be used when dyeing with certain direct and acid dyes. Which dyes are they? Explain why?

Azoic Dyes (Naphthols)

1. Why is the naphthol derivative becomes water-soluble when treated with sodium hydroxide?
2. How is an aromatic amine prepared so that it will couple with a naphthol?
3. What is the relationship between a fast color salt and a fast color base?
4. Why are the azoic dyes more durable to laundering than the direct dyes? Give two reasons.
5. In the application of azoic dyes, many times the fibers are salt rinsed before coupling. Why?
6. What are the three different methods used in preventing the naphtholate from coming out of the fibers during coupling?
7. What is the preferred intermediate step for naphthols with a high substantivity? Explain why?
8. How do you strip an azoic dye from cellulosic fibers?
9. Why is beta naphthol itself not used as a naphthol in batch dyeing with azoic dyes? Why can you use beta naphthol in continuous dyeing with azoic dyes?
10. Why are sequestering agents often used when dyeing with azoic dyes?

Vat Dyes

1. What are the chemical changes that a vat dye undergoes during its application and fixation?
2. In what form is a vat dye substantive to cellulose?
3. Why do the IN vat dyes exhibit a fast strike?
4. Why is salt used when dyeing with IK vat dyes?
5. When dyeing with vat dyes what should the dyer take care of in order to prevent over-reduction?
6. When preparing the dye-bath, the base is added before the reducing agent. Why?
7. Soap at the boil of vat dyes produces the correct shade which many times is appreciably different from the unsoaped shade. Explain why?
8. Why is it difficult to strip vat dyes from cellulosic fibers?
9. What are the advantages of dyeing fabrics with vat dyes by the <u>pigment padding</u> method?
10. Explain why indigo :
    a. is suitable for use as a pigment in solution dyeing?
    b. has a deep blue color (nature of the chromogen)?
    c. is sensitive to oxidation?
11. What are the Soluble Vat Dyes, and how are they oxidized?
12. Why is it much easier to apply the soluble vat dyes as compared to the vat dyes?
13. Explain why the soluble vat dyes do not tend to strike as fast as the regular (IN) vat dyes do?
14. Why are sequestering agents often used when dyeing with vat dyes?
15. Why are the soluble vat dyes preferred over the regular vat dyes for dyeing wool?
16. Why are the soluble vat dyes oxidized in a strongly acidic solution?
17. How are vat dyes made to be used as soluble vat dyes?
18. Why is sodium dichromate applied in presence of an acid whether it is used as an oxidizing agent or to improve wash-fastness of certain dyes?

Sulfur Dyes

1. Why are sulfur dyes relatively inexpensive?
2. Why are sulfur dyes usually available only in dull shades?
3. Why are they called sulfur dyes?
4. At present, through what range of colors do the sulfur dyes extend?
5. In exhaust dyeing, sulfur dyes are applied similar to direct dyes. Explain how?
6. What are the soluble sulfur dyes, and why are they easy to use?
7. Sulfur dyes are supplied mainly as soluble sulfur dyes. Explain why?
8. Why is it important to make sure that the pH of the bath is basic before adding the soluble sulfur dyes?
9. When dyeing with sulfur dyes, why are they given, at the end of the application, a final rinse with an alkali solution?
10. After the application of sulfur dyes, they are fixed with a mild oxidizing agent. What is the main reason for not using a strong oxidizing agent?
11. How would you strip sulfur dyes from cellulose?

Reactive Dyes

1. What makes the reactive dyes altogether different from other dyes used for cellulosic fibers?
2. What are the essential functional groups in a molecule of a reactive dye?
3. When applying reactive dyes, how is wastage of dyes minimized? Explain.
4. When dyeing with reactive dyes, the amount of salt used is much larger than the amount used when dyeing with direct dyes. Explain why?
5. When dyeing with reactive dyes, a fast addition of the base may cause a fast strike. Explain why? Give two reasons.
6. Explain why direct dyes are made to have a high substantivity while reactive dyes are made to have a low substantivity?
7. Why is dyeing with reactive dyes is relatively expensive? Give two reasons.
8. Why do reactive dyes have a relatively large number of sulfonic acid groups?
9. Explain why reactive dyes:
    a. have good leveling properties?
    b. have good water solubility?
    c. do not undergo a change in color during dyeing?
    d. are expensive?
10. How can you make a reactive dye with a bright green color by using only azo groups as chromogens?
11. Trichlorotriazine can react with amines, alcohols, and water. Show how these chemical reactions take place during the manufacturing of reactive dyes, and during their application to cellulosic fibers.
12 Even at low levels of exhaustion, most of the reactive dye reacts with the cellulose rather than with the water. What are the two main factors that lead to this result?
13. When dyeing cotton, what type of dyes would you consider first if your main concern(s) were:
    a. durability to laundering and fastness to chlorine

b. bright colors and very good wash-fastness
   c. inexpensive dyeing with very good wash-fastness
   d. red colors with very good wash-fastness
   e. very good light-fastness
   f. level dyeing
14. What makes the dye durable to washing and wet processing when dyeing cellulosics with each of the following dyes:
   a. direct   b. azoic   c. sulfur   d. vat   e. reactive
15. What are the color limitations of each of the dyes (if there are no limitations indicate so) in question #14.

Disperse Dyes

1. Are the disperse dyes water-soluble? Explain.
2. The first disperse dyes were of the 'low-energy' type. Explain why?
3. What type of dispersing agents are used with disperse dyes? At what stage are they added to the dye-bath?
4. What is the advantage of using the 'low-energy' disperse dyes over the 'high-energy' disperse dyes?
5. When are the high-energy disperse dyes preferred and why?
6. Why can't we use high-energy disperse dyes for dyeing acetate fibers?
7. Why is it more difficult to dye triacetate fibers than acetate fibers?
8. When are dye-carriers used in the dyeing of triacetate fibers?
9. How are low-energy disperse dyes chemically modified by dye manufacturers to be used as high-energy disperse dyes?
10. Can a disperse dye be used as a pigment? Explain.
11. What is the chemical nature of anthraquinone disperse dyes as compared to anthraquinone pigments?
12. To dye black a proper mixture of disperse dyes is used. Why is it that a black disperse dye is not available?
13. When dyeing with disperse dyes, what could be the effect of a strong electrolyte on the dye-bath?

Dyeing Polyester

1. How does the molecular structure of polyester explain its very low moisture content?
2. Why is it more difficult to dye polyester that other common fibers?
3. In the dyeing of polyester, why do dye-carriers promote the exhaustion of disperse dyes? Consider:
   a. the effect on the fibers   b. the effect on the dye-bath
4. What are the problems involved in using dye-carriers?

5. When dyeing polyester with disperse dyes, how are dye-carriers used:
   a. as leveling agents   b. as stripping agents
6. How does heat-setting of polyester fabrics (before dyeing) affect their dyeability? Consider heat-setting temperatures in the range of 350-420°F.

7. Explain how reduction clearing removes unfixed disperse dyes from polyester. Consider: a. azo dyes   b. anthraquinone dyes
8. When would you prefer to dye polyester with low-energy disperse dyes?
9. What type of disperse dyes would you choose for the following:
    a. atmospheric dyeing of 100% polyester that was already heat-set
    b. dyeing 100% polyester by the thermosol process
    c. dyeing polyester/cotton by the thermosol process
10. Consider the following three routes for processing a polyester fabric:
    1. Scour / Heat-Set / Dye
    2. Heat-Set / Scour / Dye
    3. Scour / Dye / Heat-Set

    Which of these sequences would you consider first (if more than one sequence is proper, indicate so) and <u>why?</u> if your main concern were:
    a. low cost
    b. over-all best results
    c. a soft hand
    d. complete removal of dye-carriers
    e. using low energy disperse dyes
    f. level dyeing (heat-setting equipment)
    g. dimensional stability of the fabric
    h. good dyeability of the fibers
    i. atmospheric dyeing of dark shades
    j. dyeing under pressure on a jet

Dyeing Nylon Fibers

1. Explain why nylon can be dyed with both soluble or disperse dyes?
2. How does the application of acid dyes to nylon differ from that of wool?
3. When you consider dyeing nylon with disperse or acid dyes:
    a. which of the two covers barre better, and why?
    b. which one gives better wash-fastness, and why?
    c. with which one is it easier to build heavy shades, and why?
4. What are the bonding sites in nylon for each of the following dyes, and what type of bonds are formed in each case (if no bonding occurs, indicate so):
    a. acid   b. disperse   c. reactive   d. basic   e. 2:1 premetallized
5. When dyeing nylon, what type of dyes would you consider first if your main concern were:
    a. level dyeing   b. durability to washing   c. deep shades
    d. brilliant colors   e. covering barre
6. Unset nylon fabric should not be dyed at the boil in the rope form. Why?
7. What may happen when heat-setting nylon with dry air at high temperatures?
8. What is the main purpose for producing basic-dyeable nylon?
9. Why is it possible to modify nylon to become either acid-dyeable or basic-dyeable?

Dyeing Acrylic Fibers

1. Why are basic dyes used extensively for dyeing acrylic fibers?
2. Why are basic dyes also called cationic dyes?
3. Cationic dyes on basic-dyeable acrylics have excellent wash-fastness. Why?
4. What is the limitation of using disperse dyes for acrylics?
5. When dyeing acrylic fibers with cationic dyes, why is it so difficult to obtain level dyeing? Give two reasons.
6. What kind of change takes place at the glass transition temperature of acrylic fibers? How does this change affect the fibers' dyeability?
7. Why are retarders used when dyeing acrylic fibers with basic dyes?
8. When dyeing acrylic fibers with basic dyes, why is a cationic retarder referred to as a colorless dye?
9. When dyeing acrylic fibers with basic dyes, how do cationic polymers act as retarders?
10. When dyeing acrylic fibers with cationic dyes:
    a. what is the disadvantage of using anionic retarders?
    b. what is the disadvantage of using cationic retarders?
11. Before dyeing acrylics with cationic dyes, what is the information needed regarding:
    a. the nature of the fibers (two characteristics)?
    b. the nature of the dyes (two characteristics)?
12. What would be the proper K value of a cationic retarder? Why?
13. What type of cationic dyes (classified by their K values) would you prefer to use to obtain light shades on acrylics, and what type would you use for dark (heavy) shades? Explain in each case why?

Dyeing Polyester/Cotton Blends

1. What are the steps involved in dyeing polyester/cotton blends with disperse and reactive dyes, by a two-bath method? What are the main advantages of this method?
2. What is the main advantage of dyeing polyester/cotton with disperse and reactive dyes, by the 'Inverse' method (dyeing the cotton first)?
3. What are the problems involved with dyeing polyester/cotton by a single dye-bath method?
4. In the continuous dyeing of polyester/cotton blends with disperse and vat dyes:
    a. What is the proper moisture content that the fabric should have before entering the dyeing range? What would be the problem if the fabric were excessively dried? What would be the problem if the fabric were too wet?
    b. Why is it impossible to apply all dyes and dyeing assistants from a single bath?
    c. Why are the IN vat dyes preferred? Give two reasons!
    d. How is migration of dyes during the drying stage is minimized? (Consider the dye-bath and the equipment)
    e. What is the content of the dye-bath
    f. What type of disperse dyes are used? Why?
    g. What prevent the vat dyes from going into the polyester during the thermosol stage?
    h. What is the purpose of passing the fabric through cooling cans immediately after thermosoling? Give two reasons.
    i. What are the proper wet pick-ups for the dye-pad and the chemical-pad? Explain.
    j. What is the content of the chemical-pad?
    k. What happens in the steamer? (consider: the vat and disperse dyes, and the fibers)

l. How are the unfixed disperse dyes removed in this process?
Consider: a. azo dyes   b. anthraquinone dyes
5. In the continuous dyeing of polyester/cotton with disperse and reactive dyes, why is salt added to the chemical-pad?

Chemical constitution of Dyes and Pigments

1. How is phthalocyanine modified to be used as:
   a. a direct dye
   b. a vat dye
   c. a reactive dye
   d. a pigment
   e. a bluish-green pigment
2. Explain why phthalocyanine:
   a. is a very stable compound
   b. has an intense bright color
3. Explain why certain vat dyes can also be used as pigments
4. Amino groups on disperse dyes serve two purposes. What are they?
5. Compare the structure of an anthraquinone disperse dye with that of an anthraquinone vat dye. Can an anthraquinone disperse dye serve as a vat dye? Explain. Can an anthraquinone vat dye serve as a disperse dye? Explain.
6. What are the three main atmospheric pollutants that cause fading of certain dyes?
7. What type of dyes show sensitivity to oxides of nitrogen? How are these dyes modified to become more stable?
8. Some disperse dyes show a high tendency to fade when they are on acetate but not on nylon. Explain why?
9. Why are tri-aryl methane sensitive to sunlight?
10. Why is indigo sensitive to sunlight?
11. Explain why alpha naphthol derivatives are rarely used in the manufacturing of azo dye?
12. The compounds I to VIII are examples of dyes and pigments.
    a. Classify these compounds according to their solubility in water.
    b. Classify these compounds according to their chromogen.
    c. Classify these compounds according to their method of application acid dyes, vat dyes, pigments, etc.). If a certain dye can be applied by more than one method indicate so.
    d. Which ones have the most intense colors? Explain in each case why?
    e. Which one(s) can be used as a vat dye as well as a pigment?
    f. Which ones are the most sensitive to sunlight? Why?
    g. In what form is indigo (V) soluble? Why?
    h. In what form is indigo colorless? Why?

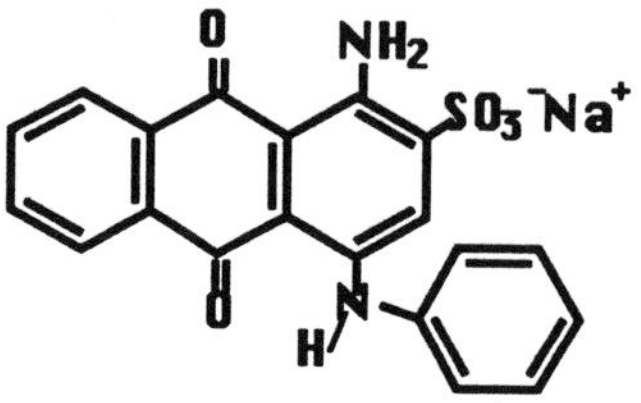

I

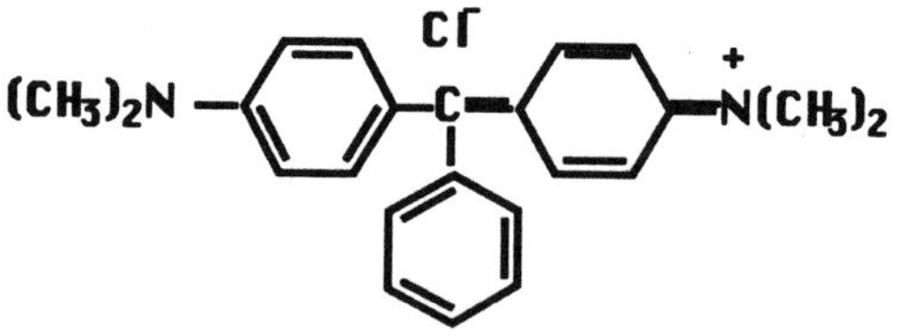

II

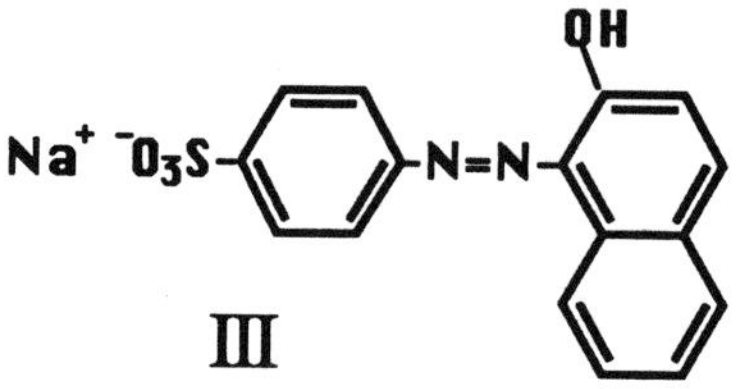

III

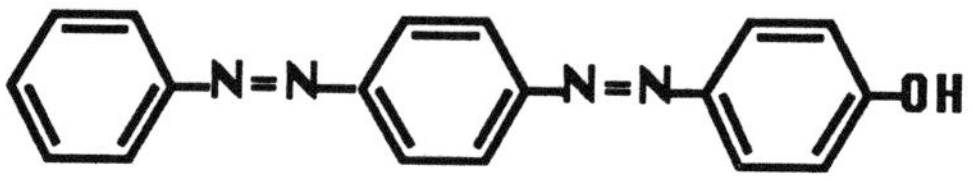

IV

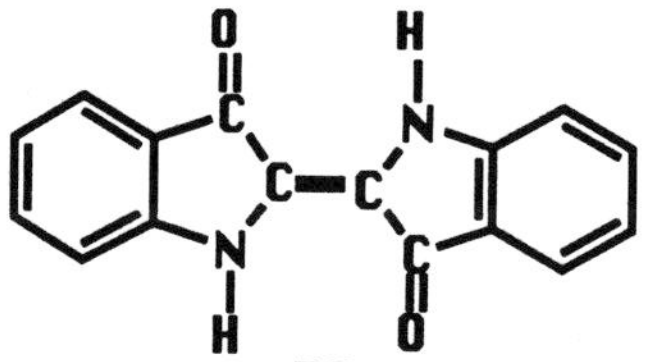

V

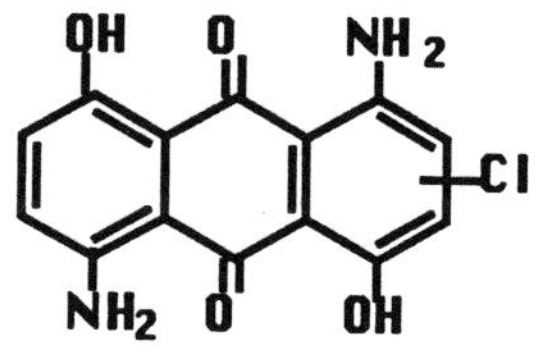

VI

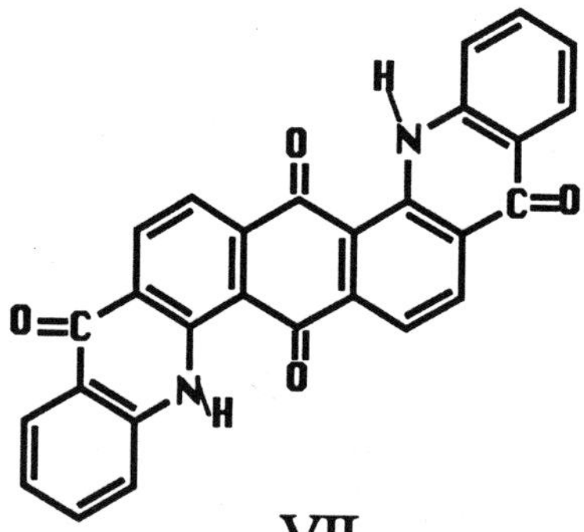

VII

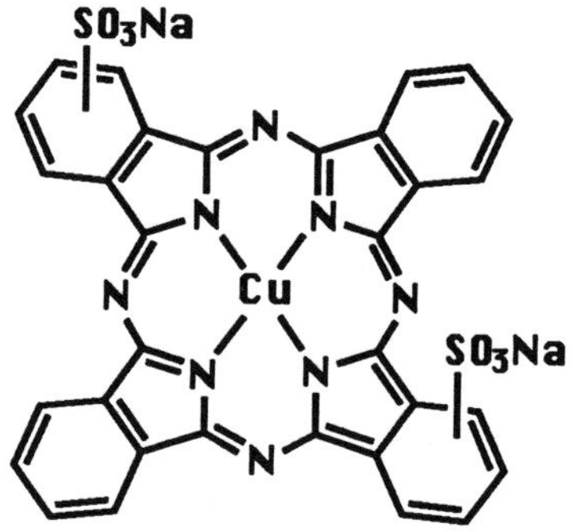

VIII

REFERENCES

Books

1. Abrahart, E.N., Dyes and their Intermediates,1977,Chemical Publishing NewYork.
2. Allen R.L.M., Studies in Modern Chemistry, 1971, Thomas Nelson and Sons Ltd. London
3. Bird C.L. and Boston W.S. (Editors), The Theory of coloration of Textiles,1975,Dyers Company Publication Trust, West Yorkshire, England. Second Edition, Editor: Johnson Alan, 1989
4. Color Index, Third Edition, Volumes 1-8, 1987, The Society of Dyers and Colorists, with acknowledgment to the American Association of Textile Chemists and Colorists, for its contribution to technical information.
5. The Dyer's World-1980's: Theory to Practice, International Dyeing Symposium, Book of papers, 1980, A.AT.C.C.
6. Dyeing and Finishing of Polyester Fibers, Manual BASF
7. Datye, K.V, and Vaidya, A.A., Chemical Processing of Synthetic Fibers and Blends, 1984, John JWiley & Sons, New York
8. Giles C.H. A Laboratory Course in Dyeing, 4th Edition 1989,D.G.Duff and R.S. Sinclair,The Society of Dyers and Colourists, Bradford Yorkshire, England.
9. Indanthren Dyes (Vat Dyes), Technical Information, BASF.
10. Nunn D.M. (Editor), The Dyeing of Synthetic-polymer and Acetate Fibers, 1979,S.D.C., Dyers' Company Publications Trust,West Yorkshire, England.
11. Peters, R.H, Textile Chemistry, Vol. III, The Physical Chemistry of Dyeing, 1975, Elsevier Scientific, Oxford, New York.
12. Kirk-Othmer, Encyclopedia of Chemical Technology, Vol 8, 1979, Wiley Interscience, New York.
13. Kukarni C.D. et al, Textile Dyeing Operations: Chemistry, Equipment, Procedure, and Environmental Aspects, 1986, Noyes Publications, New Jersey.
14. Preston Clifford (Editor), The Dyeing of Cellulosic Fibers, 1986, S.D.C., Dyers' Company Publications Trust, West Yorkshire, England.
15. Rys P. and Zollinger H., Fundamentals of the Chemistry and Application of Dyes, 1972, Wiley-Interscience, New York.
16. Sandoz LTD. Polyester Finishing, Basle, Switzerland.

17. Shore John (Editor) Colorants (Volume 1), 1990, The Society of Dyers and Colourists Publications, England.
18. Shore John (Editor), Auxilliaries (Volume 2), 1990, The Society of Dyers and Colourists, Publications, England.
19. Trotman E.R., Dyeing and Chemical Technology of Textile Fibers, Sixth Edition, 1984, John Wiley & Sons, New York.
20. Valko E.I. (Contributor), Polymer Science and Technology, Vol. 5., P235-311, John Wiley & Sons, Inc. 1964.
21. Venkataraman K. The Chemistry of Synthetic Dyes, Vol. I-VIII, 1978, Academic Press, New York.
22. Vickerstaff T., The Physical Chemistry of Dyeing, 1954, Interscience, New York.
23. Waring D R. and Hallas Geoffrey (Editors), The Chemistry and Application of dyes, Topics in Applied Chemistry, 1990, Plenum Publishing Corporation, New York.
24. Zollinger Heinrich, Color Chemistry, Syntheses, Properties and Applications of Organic Dyes and Pigments, 1987, VCH Publishers, New York.

Periodicals

-American Dyestuff Reporter, Secaucus, New Jersey

-Book of Papers, A.A.T.C.C., International Conference & Exhibition

-International Textile Bulletin (ITS), Dyeing/Printing/Finishing International Textile Service Ltd., Zurich/Switzerland

-Journal of the Society of Dyers and Colourists, Bradford, West York Shire, England.

-Review of Progress in Coloration, The Society of Dyers andColourists, Bradford, England.

-Textile Chemist and Colorist, American Association of Textile Chemists and Colorists, Research Triangle Park, North Carolina.

-Textile Research Journal, TRI/Princeton, New Jersey.

INDEX

Absorption isotherms 15
Absorption spectrum
    of azo compounds 48
Acetate
    dyeing of 155
    effect of pH 153
    properties of 152
Acetate dyes 151
Acetic acid
    pH of 199
Acid dyes 87
    application of 88
    classification of 90
    dyeing procedure 88
    mechanism of dyeing 89
    stripping of 98
Acid leveling dyes 92
Acid milling dyes 92
Acid-dyeable acrylic 179
Acrylic fibers
    dyeing with cationic dyes 179
    dyeing with disperse dyes 179
    fiber characteristics 182
    properties of 177
Acrylic/Wool Blends
    dyeing of 191
Adipic acid 170
Alkyl-trimethyl-ammonium phosphate 108
Alpha amylasa 78
Ammonium acetate 94
Ammonium chloride
    pH of 199
Ammonium hydroxide
    pH of 199
Amphoteric surfactants 26
Anhydro-glucose 100
Anionic detergents 25
Anionic surfactants 25
Anthraquinone dyes
    molar absorptivity of 49
Anti-migrant 188
Antiprecipitants 26
Antistatic agents 27
Auxochromes 47
Azo dyes
    molar absorptivity of 49
Azoic dyes
    after-treatment 116
    application of 114
    coupling reaction 111, 115
    direct printing 117
    discharge printing with 117
    intermediate step 115
    naphtholation 114
    properties of 112
Azoic dyes printing with 117
Azoic dyes, 111

Basic cyanine dyes
    molar absorptivity of 49
Basic Dyes
    see Cationic dyes 53
Basic Triarylmethane dyes
    molar absorptivity of 49
Basic-dyeable acrylic 179
Benzidine 101
    toxicity of 101
Biphenyl 154
Bleaching
    Hydrogen Peroxide 74
Bleaching agents 73
    Sodium chlorite 74
    sodium hypochlorite 74
Butyl benzoate 154

C.I. Acid
Black 60	96
Blue 158	96
Blue 25	92
Green 25	93
Orange 7	87, 92
Red 1	49
Red 85	93

C.I. Basic
Blue 3	180
Green 4	53, 180
Red 18	181
Violet 16	181

C.I. Direct
Black 22	101
Blue 218	101, 102
Blue 86	52, 102
Green 26	103, 104
Green 28	103, 104
Red 81	101

C.I. Disperse
Blue 3	175
Blue 35	162
Blue 56	150, 162
Blue 60	162, 164
Blue 7	175
Blue 73	152, 162, 164
Blue 79	152, 164
Brown 1	164
Orange 25	149, 151, 162
Orange 25	151
Orange 29	164

Orange 3   149
Orange 30   151, 164
Red 1   175
Red 11   175
Red 50   162
Red 60   162, 164
Red 65   149, 151, 162, 164
Red 73   162, 164
Red 91   162
Yellow 23   150, 162
Yellow 42   150, 162, 164
Yellow 54   162, 164, 175
Yellow 64   162, 164, 175
C.I. Fluorescent
   Brightening agent 28   55
C.I. Mordant
   Black 11   94
C.I. Pigment
   Blue 15   52
   Green 7   52
C.I. Reactive
   Blue 5   136
   Blue 7   52
C.I. Solubilized Vat
   Blue 1   129
C.I. Solvent
   Orange 9   149
C.I. Vat
   Black 25   125
   Blue 1   120
   Blue 29   52
   Blue 4   120, 122
   Green 1   125
C.I.E.
   L*a*b* color space. 44
Caprolactam 171
Carbon black 32
Carbonic acid
   pH of 199
Carbonizing 73
Cationic detergents 25
   quaternary salts 25
Cationic dyes 53, 179
   application of 181
   cationic polymeric retarders 183
   cationic retarders for 183
   characteristics 53
   characteristics of 182
   chemical nature of 54
   anionic retarders 183
   resonance structures of 53
   retarders for 183
Cationic fixing agents 108
Cationic retarders
   polymeric 183
Caustic Scouring 79
Caustic soda
   pH of 199

Cellulose 99
Cellulose acetate
   see Acetate 152
Cellulosic fibers 33
Chelating Agent 29
Chlorophylls 51
Chroma 43
Chrome dyes 94
   top chrome method 95
chromogen 47, 48
Chromophores 47
CIE system 44
classification of dyes 5
Color 37
   Additive process of mixing 41
   and chemical constitution 44
   description of 42
   Matching 42
   Subtractive process of mixing 41
   what is color? 37
Color cards 5
Color Index 5
Congo Red 101
Conjugated double bonds
   U.V. absorption 46
Continuous dyeing 2
   azoic dyes 116
   cold pad-batch method 143
   dye-selection 68
   polyester/cotton 187
   reactive dyes 143
   sulfur dyes 134
   vat dyes 127
Copper compounds
   ecological effects 109
Cotton
   caustic scouring 79
   preparation treatments 76
Cotton fabrics 76
Cotton wax 76
Cotton/wool blends 102
Coupling reaction 111
Crabbing 85
Cross-dyeing 185
Cu-phthalocyanine 49
Cyanuric-chloride 138
Cystine linkages 85

Dead cotton 82
Defoamers 166
Delocalized electrons 45
Delustering agents 42
Desizing 78
Detergents 18
   amphoteric 26
   Anionic 25
   Cationic 25
   Classification of 24

cloud point of 26
Non-Ionic 25
use of in dyeing 27
Diamond
U.V. absorption 45
Diazo component 112
Diazonium salt 111
Diazotization 111
Dibromo-indigo 121
Diphenylpolyenes 46
Direct dyes 101, 103
    after-treatments of 108
    application of 104
    classification of 107
    copper complex of 101
    effect of electrolytes 105
    effect of pH 107
    effect of solubility 105
    effect of temperature 106
    late exhaustion of 107
    maximum exhaustion temperature 106
    properties of 103
    stripping of 109
    structure of 101
Discharge printing 117
Disperse 162, 164
Disperse dyes 146
    application of 148
    chemical characteristics of 149
    classification of 150
    commercial forms of 147
    high energy 151
    low energy 151
    particle size 148
    rate of dyeing 148
    reduction clearing 163
    solubility 146, 148
    volatility of 150
Dispersion 18
DTPA 28, 165
Dye molecules
    chemical structure of 47
    relative size 197
Dye-carriers 153
Dye-classification 30
Dye-selection 194
    main considerations 194
    other considerations 194
Dyeing
    .i.Exhaust dyeing
        mechanism 11
Dyeing equipment 57
    .i.Continuous dyeing
        equipment 66
    Beck Dyeing Machine 61
    exhaust dyeing, 57
    Fabrics 61
    Jet Dyeing Machines 63

    Jig Dyeing Machine 62
    Package Dyeing Machines 59
    Raw Stock 57
    Yarns 58
Dyes
    characteristics of 3
    Chemical Structures of 16, 47
    Classification According to Chemical Constitution 30
    Classification According to Method of Application 30
    Classification According to Solubility 30
    information on 5
    Physical Forms of 35
Dynamic equilibrium
    in exhaust dyeing 13

EDTA 28, 123, 165
    Stripping of direct dyes 109
Electromagnetic spectrum 37
Electron acceptors
    chromophores 47
Electron donors
    auxochromes 47
Emulsion 18
Enzymatic Desizing 78
Enzymes
    alpha amylasa 78
EPA 193
Equalizing dyes 92
Ethylene glycol 17
Exhaust dyeing 1, 8
    absorption isotherms 14
    after-treatment 13
    dye solution 10
    effect of temperature 13
    fixation 13
    form of substrate 10
    major steps 8
    mechanism 11
    preparation 9
    the dye-bath 9

Fast strike 12
Fastness ratings 196
Fastness requirements 196
Fastness to crocking 196
Fats
    basic hydrolysis of 23
Fatty acids 22
Faulty dyeing 195
Fiber Classification 33
Fibers
    amorphous regions 6
    crystalline regions 6
    glass form 7
    moisture regain of 34
    rubber form. 7

Fillers 35
    for direct dyes, 35
    for disperse dyes 35
Fluorescent brightening agents 54
Formaldehyde derivatives
    toxicity of 109
Formic acid
    pH of 199
Freundlich isotherm 15

Glass transition temperature 7, 82
Glauber salt
    pH of 199
Glycerol, 17, 22
Graphite 32

Hard water 23, 24, 27
Heat-Setting 82
HEDTA 28, 165
Heme 51
Hexamethylene-diamine 170
Hue, 43
Hydro 122
Hydrochloric acid
    pH of 199
Hydrogen-peroxide 74
Hydrophilic fibers 33
Hydrophobic Fibers 33
Hydrosulphite 122

Inappropriate dyeing 195
Indanthrone 120
Indigo 120
Indigo, 32
Indigo-fera plants 121

Jet dyeing machines 63
Jig Dyeing Machine 62

Keratin 84
Kevlar 171

L*a*b* color space 44
Langmuir isotherm 15
Late exhaustion 107
Leuco-indigo 122
Level dyeing 2
Light-fastness 196
Liquor ratio 57
Light scattering 42

Mass coloration 1, 32, 50, 55
Mauve. 53
Mercerization 81
Metal complex dyes 96
Metamerism 42
Methacrylic acid 177, 178
Methyl-methacrylate 177

Methyl-naphthalene 154
Micelle 21
Milling 87
Milling dyes 92
Mixing colors 40
Modacrylic fibers 33
Modacrylics 177
Moisture regain of fibers 34
molar absorptivity 3, 44
    of Chromogens 49
Molar extinction coefficient, 44
Mordant dyes 94
Motes 76
Munsell system 4
3
Muriatic acid 199
Naphthalene sulphonic acid condensates 147
Naphtholate 111
Naphtholation 111
Naphthols 111
Natural pigments 9
Nernst isotherm 15
Neutral dyeing acid dys 93
Nitrogen oxides 155
Nitrophenyl amine derivatives 150
Nomex 171
Non-ionic detergents 26
    cloud point 165
Nylon 170
    acid-dyeable 176
    basic dyeable nylon 176
    deep and light-dyeable 176
    dye selection 172
    dyeing with Metallized Dyes 174
    dyeing with acid dyes 173
    dyeing with acid leveling dyes 174
    dyeing with disperse dyes 175
    dyeing with metallized dyes 173
    heat-setting of 172
    modified fibers 175
    nylon 6 170
    nylon 6,6 170
    properties of 170
Nylon / wool blends
    dyeing of 189

O-phenyl-phenol 154
Olefins 33
Optical brighteners
    application of 55
    chemical
    structure of 55
OSHA 193

Palmitic acid 22
Pectins 79
Perchloroethylene 18
pH Scale 198

pH values of chemicals 199
Phosphoric acid
    pH of 199
Phthalocyanine 50
    chemical structure 50
    cobalt complex 50
    copper phthalocyanine 50
    Dyes 49
    Pigments 49
Pigment dyeing 32, 50
Pigment printing 32, 50
Pigments 31, 32
Polyacrylonitrile 33
Polyamides 33
Polyester 157
    atmospheric dyeing 161
    basic dyeable 158
    continuous dyeing of 168
    dye selection 162, 164
    dyeing procedures for 159
    heat-setting of 158
    oligomers 160
    pressure dyeing 163
    properties of 158
    rapid dyeing under pressure 164
    reduction clearing 162
    sequence of processing 160
    stripping disperse dyes from 169
    thermo-migration 159
    transfer printing of 169
Polyester/cotton blends
    Continuous Dyeing of 187
    Continuous Dyeing Procedures 187
    dyeing with Disperse and Reactive Dyes 189
    Exhaust Dyeing of 185
    with Disperse and Vat Dyes 187
Polyester/wool blends
    dyeing of 190
Polyethoxy chain 25, 26
Polyethylene teraphthalate 157
Polyphosphates 29
Polyurethanes 33
Polyvinyl alcohol 78
    lubricating agent 78
    partially hydrolyzed 78
Potassium dichromate 95, 124
Potassium persulfate 178
Premetallized acid dyes 96
Preparation treatments 72
    wool 73
Printing
    with azoic dyes 117
Protein fibers 33

Rayon 99
    polysonic fibers 100
    structure of 99
Reactive dyes 136
    alkali additions 141
    chemical structure 136
    cold pad-batch method 142
    continuous dyeing with 143
    dye substantivity 140
    dyeing temperature 142
    exhaust dyeing 139
    hetero-bifunctional dyes 144
    Liquor Ratio 141
    Procion dyes 138
    properties of 139
    reaction with cellulose 137
    Reaction with water 137
    salt additions 141
    Vinyl Sulfone type 138
Reduction clearing 163
Ring-dyeing 3

Saponification 22
Sequestering Agents 28
Shade cards 5
Shade reproducibility 193
Singeing 77
Soap 18, 22
    limitations 23
    making of 22
Soda ash
    pH of 199
Sodium acetate
    pH of 199
Sodium bicarbonate
    pH of 199
Sodium carbonate
    pH of 199
Sodium chlorite 169
sodium chlorite, 74
Sodium dichromate 95
Sodium dithionite 122
Sodium dodecylbenzene sulfonate 24
Sodium hydroxide
    pH of 199
Sodium hypochlorite 74
Sodium lauryl sulfate 24
Sodium meta silicate
    pH of 199
Sodium percarbonate 76
Sodium poly metaphosphate 29
Sodium polysulfide 131
Sodium silicate 75, 80
Sodium stearate 19
Sodium sulfate
    pH of 199
Sodium tripolyphosphate 29
Sodium- lignine sulfonate 147
Sodium-alkyl-aryl-sulfonate 25
Sodium-alkyl-sulfate 25
Sodium-alkyl-sulfonate 25
Softeners 27

Soluble Sulfur Dyes 133
Soluble vat dyes 128
Solution dyeing 1, 32, 50, 55
Solvent Dyes 33
Stilbene derivatives 55
Stripping 3
Sulfato-ethyl-sufone 139
Sulfonated styrene 178
Sulfur dyes 131
    application of 133
    chemical nature of 131
    continuous dyeing 134
    exhaust dyeing 134
    properties of 132
    synthesis 131
Sulfuric acid
    pH of 199
Sulfonated styrene 177
Super milling dyes 91, 93
Surface tension 19
Surfactants 18
Syndets 24
Synthetic Detergents 24

Tanic acid 53
Temperature conversion 200
Tetra sodium pyrophosphate 29
Thermoplastic fibers 7
Thermosetting 82
Thermosol process 168
Thioindigo 120
Thiourea-dioxide 168
Tippy dyeing 87
Titanium dioxide 42
Transfer printing 169
Transition metals
    ecological effects 109
Triacetate 152
    dyeing of 156
Trichloro-1,3,5 triazine 138
Trichlorobenzene 154
Trichloroethylene 18
Trisodium phosphate 29
Tyrian purple 121

Union dyeing 102, 185

Value 43
Vat dyes 119
    anthraquinone derivatives 120
    application of 122
    classification of 125
    continuous methods 126
    dyeing 123
    history 121
    indigoids 120
    leuco form of 119
    oxidation 124
    pigment dyeing method 126
    properties of 121
    Reduction (vatting) 122
    soluble vat dyes 128
Vinyl chloride 177
Vinylidene chloride 177
Viscose 99
Visible light 37, 44

Wash-fastness 196
Water Softening 27
Wetting agent 188
Wetting agents 27
Woad plants 121
Wool 84
    alpha helix configuration 85
    cleaning of 73
    cystine cross-links in 85
    dyeing with Acid Dyes 89
    ionic cross-links in 86
    iso-electric point of 89, 93
    macro Structure of 86
    micro Structure of 85
    primary structure of 85
    Stripping dyes from 98
    structure of 84
Wool blends
    dyeing of 189
Zinc-sulphoxylate-formaldehyde. 98